AF412869

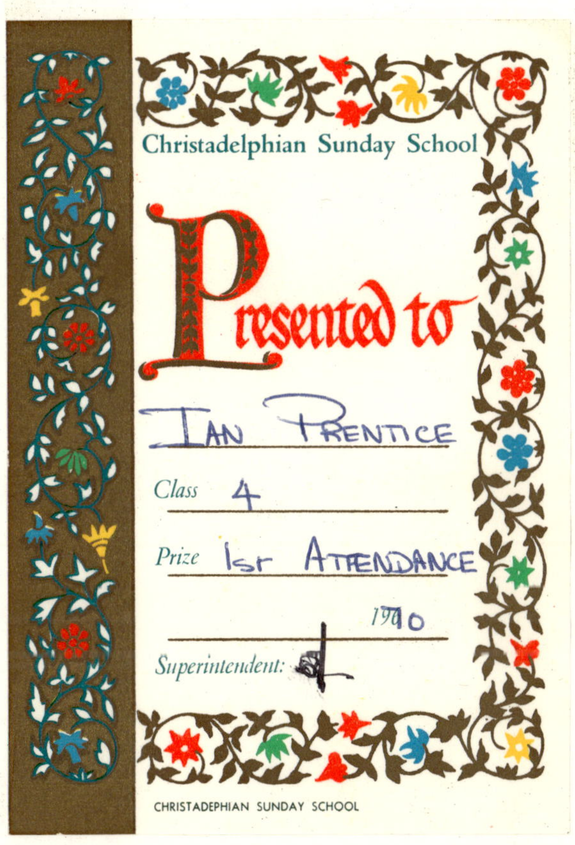

BEGINNER'S GUIDE
TO PRACTICAL ELECTRONICS

BEGINNER'S GUIDE TO PRACTICAL ELECTRONICS

Written and Illustrated by
R. H. WARRING

LONDON
LUTTERWORTH PRESS

First published 1966
Second impression 1969

1788 0033 8

COPYRIGHT © 1966 R. H. WARRING

*Printed in Great Britain by
Cox & Wyman Ltd., London,
Reading and Fakenham*

CONTENTS

Chapter		Page
1.	INTRODUCTION TO ELECTRONICS	9
2.	CIRCUIT CONSTRUCTION	20
3.	INSTRUMENTS	35
4.	ELECTROMAGNETIC COILS AND RELAYS	49
5.	VALVES AND SIMPLE VALVE CIRCUITS	59
6.	SEMICONDUCTORS (DIODES AND TRANSISTORS)	71
7.	PHOTOCELLS AND SOLAR CELLS	84
8.	PHOTOCELL CIRCUITS AND DEVICES	92
9.	TRANSFORMERS AND POWER PACKS	108
10.	AMPLIFIERS, HEARING AIDS AND INTERCOMS	124
11.	OSCILLATORS	139
12.	RADIO RECEIVERS	154
13.	RADIO CONTROL	166
14.	THE OSCILLOSCOPE	180
	GLOSSARY	189

LIST OF DESIGNS AND WORKING CIRCUITS
TO MAKE

Figure numbers are indicated in parentheses

Chapter

1. Neon circuit (1.2) — Neon flasher (1.3) — Day/Night automatic flasher (1.4) — Electrical bridge (1.5) — Electronic bridge (1.5) — Electronic resistance/capacity bridge (1.6) — Audio oscillator (1.7, 1.8, 1.9) — Speed controller for model cars, etc (1.10).

2. Simple radio receiver (2.1).

3. Increasing range of milliammeter (3.2) — Milliammeter into Voltmeter (3.3) — Multitest meter (3.4) — Transistorized "valve" voltmeter (3.5) — High sensitivity voltmeter (3.6) — Resistance substitution box (3.8, 3.9) — Capacitance substitution box (3.10) — Geiger counter probe (3.11) — Geiger probe ratemeter (3.12).

4. Air cored coils (formula) — Sensitive relay coils (4.1).

5. Diode rectifier (5.4) — Smoothed rectifier (5.5) — Double-diode Power pack (5.6) — Triode amplifier (5.8, 5.10) — Simple valve radio (5.9) — Single valve radios (5.11).

6. Transistor test circuits (6.10, 6.12) — Simple transistor tester (6.14).

7. Sun-powered radio (7.3) — Solar powered electric motor (7.4) — Simple lightmeter (7.3) — Transistor light switch (7.7) — Temperature stabilized light switch (7.8).

8. Burglar alarm (8.2) — Auto-reset burglar alarm (8.3) — Photosensitive switch (8.4, 8.5) — Annunciator relay (8.6) — Relay with "hold" (8.7) — Extra sensitive photoswitch (8.8) — Garage door opening (8.10) — Slave relay circuit (8.11) — Hold circuits (8.12) — Light-operated switch (8.13) — Light meter (8.14) — Master photocell unit (8.15).

9. Power pack transformer — Oscilloscope transformer — Photocell transformer — Basic power pack (9.7) — Rectifier circuits (9.8) — Voltage doubler (9.8) — Complete power pack (9.9) — Simple power pack (9.10) — Low voltage power pack (9.11) — Stabilized DC power pack (9.12) — Twin DC power pack (9.13).

10. Valve amplifier (10.1) — 3-watt amplifier (10.2) — Stereo amplifier (10.3) — Transistor amplifier (10.4) — Transistor driver (10.5) — 5-watt transistor amplifier (10.6) — Hearing-aid (10.7) — Sub-miniature hearing aid (10.8) — Modern hearing aid (10.9) — Batteryless intercom (10.10) — Baby alarm (10.11) — Transistorized intercom (10.13).

11. Signal generator (11.5) — AF/RF oscillator (11.6) — Light-operated oscillator (11.7) — Multivibrator (11.8) — Very low frequency vibrator (11.9) — Flashing lights (11.10) — Metronome (11.11) — Electronic organ (11.13).

12. Aerial coils (12.3, 12.4) — Crystal receiver (12.5) — Simple transistor receiver (12.6) — Improved transistor receivers (12.7, 12.8) — Reflex receiver (12.9) — Mullard 3-transistor receiver (12.10).

13. Simple radio control transmitter (13.1) — Absorption wave-meter (13.3) — Field strength meter (13.4) — Crystal controlled transmitter (13.5) — Improved transmitter (13.6) — Tone generators (13.7) — Simple radio control receiver (13.9) — Receiver with transistor amplification (13.10) — Relayless receiver (13.11) — All-transistor receivers (13.12, 13.13) — Tone receiver (13.14).

14. Simple oscilloscope (14.2).

CHAPTER 1

INTRODUCTION TO ELECTRONICS

IT is a little difficult to justify the description "electronics" as a separate entity from "electricity" since both are part and parcel of the same thing. In other words, both electrical circuits and electronic circuits are concerned with movement of charges or electrons around the circuit, and the behaviour of such charges when meeting different components in the circuit.

Simple electrical circuits, especially direct current or DC circuits, are readily understood since they work in a very logical way. When it comes to introducing components whose working behaviour is not so obvious, such as valves or transistors, the resulting circuits have come to be called "electronic". In point of fact they are still just electrical circuits—nothing more, nothing less.

The only major difference is that electronic circuits involving the use of valves or vacuum or gas tubes invariably demand the use of higher voltages than is usual with simple electrical circuits. Whilst this can be met with high tension batteries, it raises the running cost of the circuit. A better scheme is usually to derive high tension and any other necessary voltages from a power pack, connected to a normal mains supply. Because of their usefulness in this respect a separate chapter is devoted to power packs. Transistor circuits overcome this particular limitation of requiring high voltage supplies as well as usually having a low current demand and can be battery powered quite economically. Again, however, a power pack may well be utilized in cases where it is desirable to eliminate the battery, or rather where higher battery voltages are called for.

The term "electronics" is modern and thus can have a special appeal to the more knowledgeable designers and constructors of modern electrical circuits. To the beginner such a description can be off-putting since it implies something more difficult to understand or work with than basic electricity. This is really nonsense. It is just as easy to understand the behaviour of valves and transistors as it is to know how a resistor or a capacitor functions. Certainly they may do more, and have a more varied application than simple electrical components, but there is nothing all that mysterious about them. It is only necessary to understand what they do—not necessarily how they work—to be able to incorporate them in a working circuit. Even that is not necessary. Given a proven circuit design, it will work if connected up properly whether it is just a bell and battery circuit (simple electricity), or one involving valves or transistors (an electronic circuit). If it does not work, this does not mean that the constructor does not understand the function of valves or transistors, but rather that he has wired the circuit up incorrectly, or perhaps one of the components in the circuit is faulty.

Digging a little deeper to find out how valves and transistors work, however, not only makes them acceptable as commonplace circuit components, but enables the experimenter to be more adventurous. He can modify circuits to his own requirements with a knowledge of how to go about such changes and the result he is likely to achieve. He will also be in a better position to find faults which may be present in particular circuits.

The emphasis throughout this book is on *practical* electronics, and so descriptions of working principles involved have been kept to a minimum—enough, it is hoped, to enable the reader to appreciate the working of less familiar components and the reason for employing them in particular ways in particular circuits. Some of these descriptions may be a little difficult to absorb on first reading—the biasing of transistors, for example. If re-read—several times if necessary—they will suddenly make sense, especially when applied to a working circuit. The term "electronics" has then lost its mystery and starts to become just another practical subject. It is, after all, the practical aspects of electronics that are the most interesting and most rewarding to pursue.

The majority of the working circuits described in the book—and there are something like one hundred different designs—are basically very simple, even though they may employ "electronic" components. Strictly speaking, under the title, it would have been perfectly justifiable to include simple "electrical" circuits and experiments, but we have tried to avoid the too simple, and the too obvious. There are plenty of books describing simple electrical experiments, and the subject at that level is not particularly interesting. The bulk of it has probably remained basically unchanged for up to half a century or more, in fact. We hope, therefore, that the circuit designs are new to most readers—

as well as being essentially modern—and as a consequence, interesting.

Basically this is a book for beginners, so all the circuit designs and descriptions have been kept simple. Some of them may not appear simple on first reading or study, but the comment made above relative to transistors applies. The main thing is not to be put off by words. Radio control, for instance, may seem a very advanced subject which can only be understood and mastered by experts. In fact, straightforward single-channel radio control transmitters and receivers are very easy to make and get working. Almost anyone who can use a soldering iron properly can produce working circuits of this type.

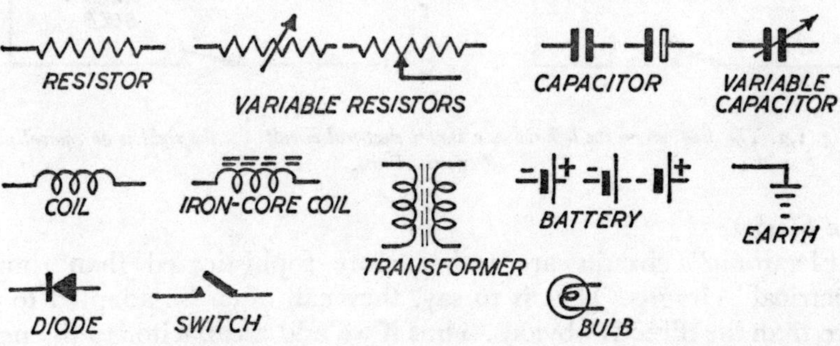

Fig. 1.1. Symbols used in circuit diagrams.

It is essential, of course, to be able to identify symbols used in circuit diagrams. The more common symbols are shown in Fig. 1.1. There are others and some of these will be introduced later in the chapters dealing specifically with valves, photocells and transistors. It is easy enough to remember these symbols, and so identify the various components involved in a circuit diagram. Visualizing them in terms of actual physical shape and size of component is not so easy, but this comes with some experience, and the explanatory notes given in Chapter 2 will form a useful guide.

Now have a look at the two circuits shown in Fig. 1.2. One is a simple electrical circuit which is completely familiar—a light bulb in a circuit with a battery and switch. An equivalent "electronic" circuit employs a neon bulb instead of an ordinary bulb, and because this is a discharge bulb rather than one which lights up merely by heating of a filament, it requires a much higher voltage. A variable resistor is also incorporated in the circuit to adjust the voltage, as necessary, for the neon to "strike" and light. Once a satisfactory value of resistance has been found the potentiometer (variable resistor) can be replaced by a fixed resistor of a value equivalent to the setting of the potentiometer.

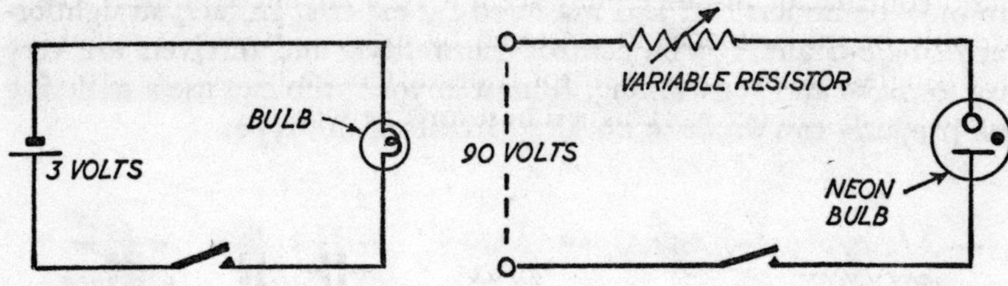

Fig. 1.2. *The diagram on the left shows a simple electrical circuit. On the right is an equivalent electronic circuit.*

Neon Flasher

"Electronic" circuits are rather more sophisticated than simple "electrical" circuits. That is to say, they can often be adapted to do more than the directly obvious. Thus if we add a capacitor to the neon circuit, as shown in Fig. 1.3, we have an automatic flasher circuit. The

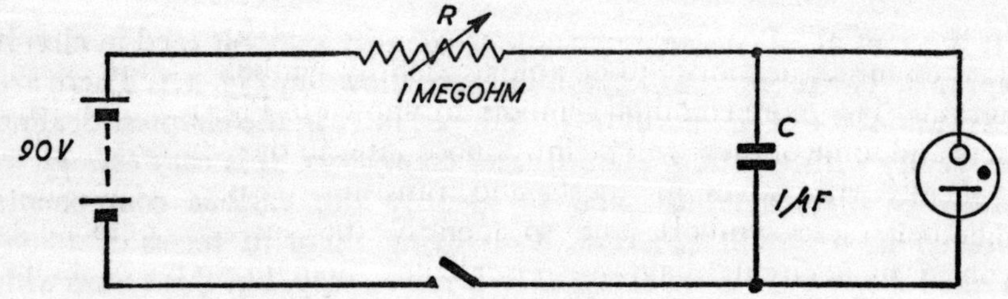

Fig. 1.3. *Automatic flasher circuit.*

variable resistor is adjusted so that there is excess resistance in circuit to limit the voltage to just below the striking voltage of the neon. However, when the circuit is completed by closing the switch the capacitor charges up and subsequently discharges. The discharge will boost the voltage across the neon, so that it lights as long as the capacitor continues to discharge. It will then go out. This cycle of the capacitor charging and discharging will be repeated, causing the neon to flash on and off. The time of flashing will depend on the values of R and C, and with the component values shown will be of the order of three-quarters of a second.

Day/night Automatic Flasher

We can add further sophistication to this very simple circuit by extending it to include a photocell, as in Fig. 1.4. When light falls on the photocell it will have a low resistance and thus bypass current from the neon, in other words more or less short out the neon so that it will not strike. When darkness falls (or the illumination of the photocell is

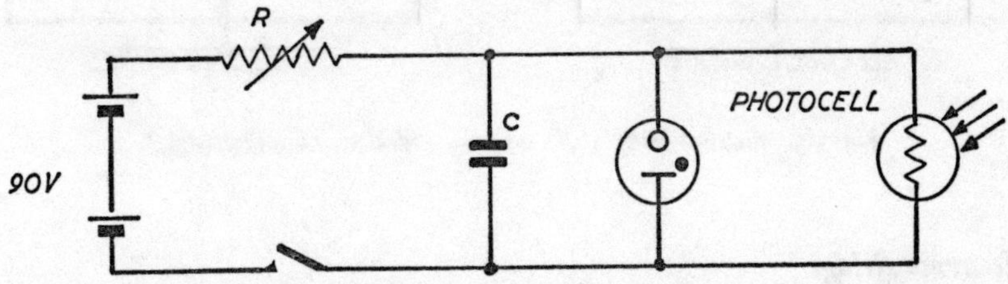

Fig. 1.4. Day/night automatic flasher circuit.

reduced for any reason) the photocell will develop a high resistance and the main flow of electricity will be through the neon circuit. This is a flasher circuit, as we have already seen. Thus the circuit of Fig. 1.4 will remain idle or non-working all the time the photocell is illuminated as in daylight for example, but automatically switch itself on to give a flashing light from the neon during the dark. There is nothing complicated about it. It just works that way, if set up properly.

Now consider a simple resistance bridge. This circuit is probably familiar from school science and comprises, basically, a resistance of known value R_k and a resistance of unknown value R_x connected to a potentiometer as in Fig. 1.5. In the electrical bridge a meter is connected in series with a battery across the bridge as shown in the first diagram. By adjusting the potentiometer so that a zero reading is obtained on the meter —i.e. balancing the bridge circuit—the value of the unknown resistance can be determined from the setting or calibration of the variable resistance RV and the known value R_k.

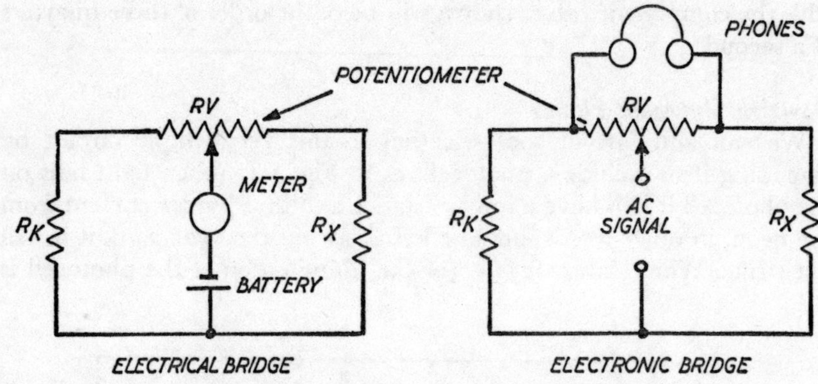

Fig. 1.5. Electrical bridge (left) compared with electronic bridge (right).

Electronic Bridge

The equivalent electronic bridge works in exactly the same way, except that instead of a meter and battery an audio frequency signal is applied across the bridge and the effect listened to via a pair of earphones connected as shown. It does the same thing as the electrical bridge in a rather more interesting way. Balance, in this case, is determined by adjusting RV until the sound heard in the phones is minimal, or disappears entirely.

A further advantage offered by the electronic bridge is that it can be used with capacitors as well as resistors to find the value of an unknown capacitor by comparing it with one of known value. This cannot be done with the simple electrical bridge since that works on DC and direct current will not flow through capacitors. The electronic bridge works on AC, and capacitors are conductive on AC to complete the bridge circuit.

The actual circuit required is a very simple one and need consist of nothing more than a potentiometer wired to three pairs of terminals, as shown in Fig. 1.6. This is a complete instrument for measuring the values of unknown resistors and capacitors, except for the source of AC signal, which we can leave for the moment. Using known resistors as the standard we can calibrate the potentiometer to read the value of unknown resistors directly. The best way to do this is by connecting the standard resistor and then a series of *known* resistors to the 'unknown' terminals, finding the balance point and marking the appropriate

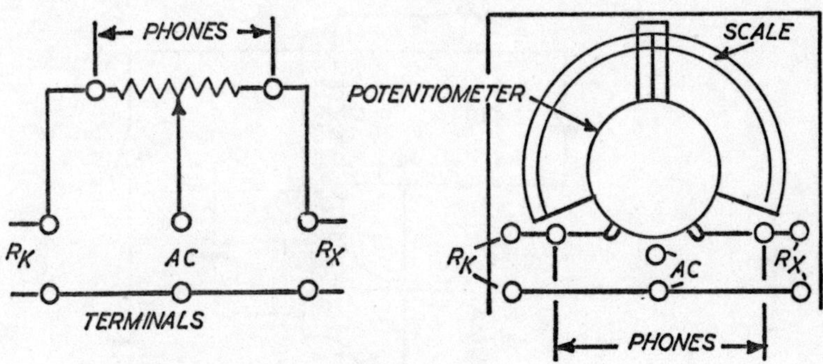

Fig. 1.6. *Electronic bridge with calibrated potentiometer.*

position on the slider scale. Using a 1 kilohm potentiometer for RV, the following ranges of resistor measurement can be calibrated:

With 1 kilohm as the standard—10 ohms to 100 kilohms

With 100 kilohms as the standard—1 kilohm to 10 megohms

Similarly with capacitors, and again using the same 1 kilohm potentiometer for RV.

With ·001 μf for the "standard"—range 10 pf to 0·1 mf

With ·1 μf for the "standard"—range ·001 mf to 10 mf

We have, in fact, produced a very useful and versatile test instrument with very little cost and effort.

Audio Oscillator

However, there is still the question of a source of audio frequency supply to be applied to the bridge. This demands the use of an audio frequency oscillator. A circuit for a suitable oscillator is shown in Fig. 1.7. This is a design by Mullard and is reproduced just as it might appear in a technical manual or specialist electronic journal. It may look terribly complicated and impossible to understand at first. In fact, it is a very simple circuit to make, as will become obvious once we translate it into terms of actual physical components. We can count in the circuit three resistors (R1, R2 and R3), two capacitors (C1 and C2), a transistor (TR1), a switch (S3) and a battery (B1). All these are standard

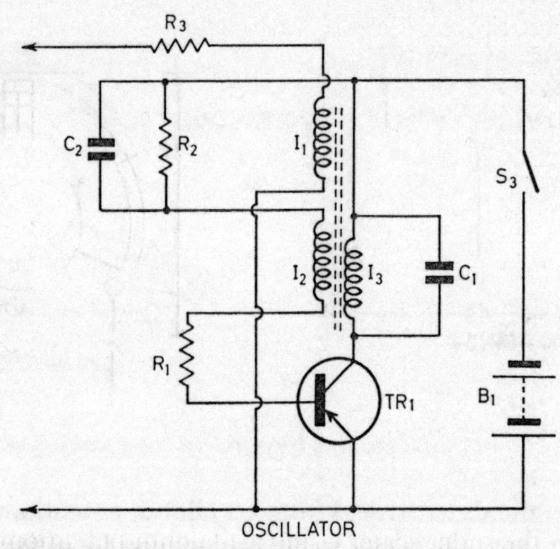

Fig. 1.7. Fixed frequency audio oscillator.

Components list:

R1	3 *kilohms*	C2	·1μ F 6 *volt wkg.*
R2	470 *kilohms*	TR1	OC71
R3	100 *ohms*	B1	4·5 *volts*
C1	·1μ F 6 *volt wkg.*		

components which can be bought from any radio dealer who specializes in supplies for the amateur constructor. This leaves the three coils (I_1, I_2 and I_3) as the unknown component. It looks a very technical problem to have to solve. In fact the circuit specification will include full details of how these coils are to be made and it boils down to nothing more than the simple coil winding job shown in Fig. 1.8. A $1\frac{1}{2}''$ length of Ferroxcube FX1104 (approximately $\frac{5}{16}''$ diameter) is fitted with six circular discs about $1\frac{1}{4}''$ diameter cut from thin sheet insulating material (thin ply will do) and wire wound on to the three bobbins made up as shown in Fig. 1.8. This forms the complete assembly for I_1, I_2 and I_3.

This coil unit can then be mounted on a Paxolin panel of suitable size (about $5'' \times 1\frac{1}{2}''$), either by cementing down or even strapping in place with a rubber band. Solder tags (or short bus-bars) can be fitted to the panel to carry the other components and the complete circuit with a 1 kilohm linear potentiometer finished to look something like Fig. 1.9. Provided all the connections are correctly made (and the only real chance of going wrong is connecting the coils up the wrong way

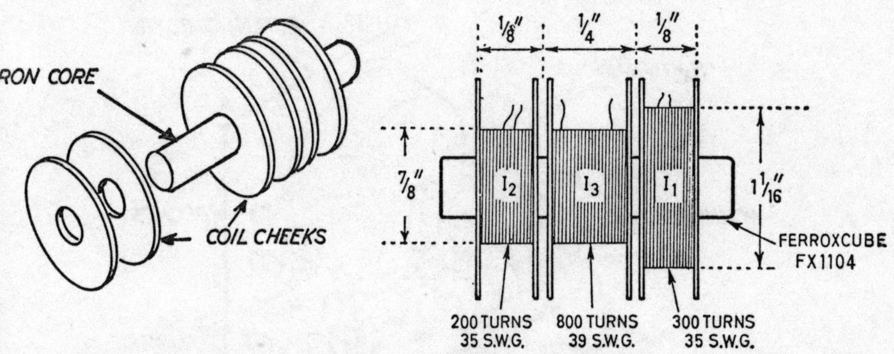

Fig. 1.8. Coil winding for audio oscillator (Fig. 1.7).

round) the circuit will work as a 6 kilocycles/second audio oscillator as soon as the battery is connected up and the circuit completed by switching on.

With very little effort, and no special knowledge, we have made and got working what appears, by its description, to be a most advanced electronics device—an audio frequency oscillator. We certainly do not need to know how it works (although it is to be hoped that Chapter 11 will even explain that in simple terms). All we need to be able to do is to be able to translate a circuit diagram into a practical circuit for assembly. Practical electronics can be just as simple as that.

Incidentally we can easily test this oscillator on its own. By connecting the output (two arrowed lines) directly to a pair of headphones and operating switch S3, a tone will be heard. This tone or note can be altered by using different values for the capacitor C2. Decreasing the value of this capacitor will produce a higher note, and vice versa. This circuit is, in fact, a simple tone signal generator.

Transistorized Potentiometer (or Voltage Controller)

This neat little device (Fig. 1.10) is another typical example of how an electronic circuit can provide a distinct advantage over a pure electrical circuit. By combining a single power transistor with a standard potentiometer a performance is obtained equivalent to that of a

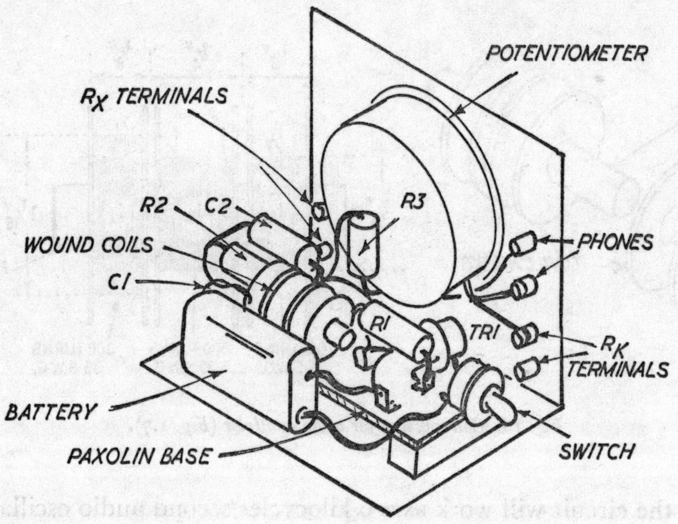

Fig. 1.9. *Audio oscillator (as Fig. 1.7) coupled to a 1 kilohm linear potentiometer.*

heavy-duty potentiometer which is usually costly to buy as well as being extremely bulky.

The output voltage from this circuit is drawn from the transistor and is infinitely variable from zero to a maximum by adjusting the base-emitter voltage of the transistor via the potentiometer RV1. The fact that the output is taken from the emitter also means that the output impedance is low (of the order of 3 ohms). The circuit is thus particularly suitable as a voltage controller for supplying low impedance devices, such as transistor amplifiers and model train or model car tracks.

An OC35 transistor will be capable of handling a maximum power of 30 watts, or 2·5 amps at 12 volts output. Because of the fairly high current values it will be called upon to carry, a heat sink is essential. To provide this the transistor should be bolted to a piece of 16 s.w.g aluminium sheet of generous area.

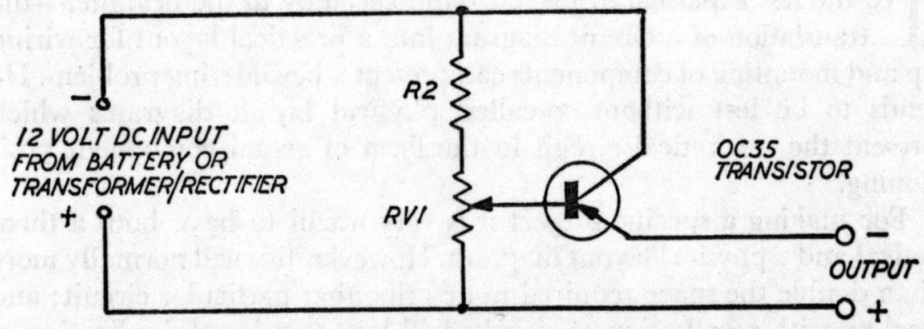

Fig. 1.10. Transistorized potentiometer.

To pass an emitter current of 2·5 amps an OC35 transistor needs a base current of 75 milliamps. For a 15 volt supply (input) the combined value of RV1 and R2 must be such that this current is available, i.e.

$$\text{resistance} = \frac{\text{volts}}{\text{amps}}$$

$$= \frac{12}{\cdot 075}$$

$$= 160 \text{ ohms.}$$

In practice the combined resistance should be considerably smaller, say one half of this value—80 ohms. The purpose of R2 is merely to act as an end resistance to the potentiometer and can be quite small—say 5 ohms. The value of RV1 required is therefore about 75 ohms maximum. Since this is not a standard size, RV1 could be a 100 ohm potentiometer. This would then make a controller suitable for any input voltage between 12 and 18 volts. For a lower voltage supply a suitable (lower) value of R2 and RV1 combined can be calculated on the same lines as above.

CHAPTER 2
CIRCUIT CONSTRUCTION

To the less experienced person—and certainly to the beginner—the translation of a circuit diagram into a practical layout for wiring up and mounting of components can present a bewildering problem. He tends to be lost without so-called physical layout diagrams which present the theoretical circuit in the form of actual component positioning.

For making a specific project it is very useful to have both a theoretical and a physical layout diagram. However, this will normally more than double the space required to describe that particular circuit; and dealing with a collection of circuits will lead to a lot of duplication of physical layout descriptions. Further, it only needs a little experience in circuit construction to know how to go about the job of circuit assembly, working from a circuit diagram, with the advantage of adapting or using materials to hand.

It is circuit designs themselves which are really of interest to the experimenter and electronics enthusiast and so to include as many different circuits and designs as possible the individual descriptions in this book are confined to circuit diagrams and a full list of components, leaving the matter of actual physical assembly to be worked out, where necessary. This chapter sets out to describe the many ways in which assembly can be tackled—from crude, simple arrangements which can be used to complete a circuit for testing in a minimum of time and effort to the professional type of assemblies necessary for the construction of permanent sets for regular use.

The greater majority of circuits described in this book are quite simple and do not involve a multiplicity of components requiring wiring up and/or mounting. As far as possible, too, the circuit or theoretical diagrams have been laid out logically so as to present a picture of the circuit layout required. That is to say, the physical layout of components should, as far as possible, follow the same positions shown on the circuit diagram.

First step in deciding on a suitable layout, therefore, is to consider the circuit with actual component shapes replacing the symbols, as in Fig. 2.1. This will indicate which components need mounting or fixing to some sort of rigid base and which can be wired directly to other components. It will also indicate where crossing leads or bridging wires will probably have to be used. Having transformed the circuit diagram into some sort of physical picture it is then a matter of considering how to construct that picture in its final form—the manner of going about the circuit assembly.

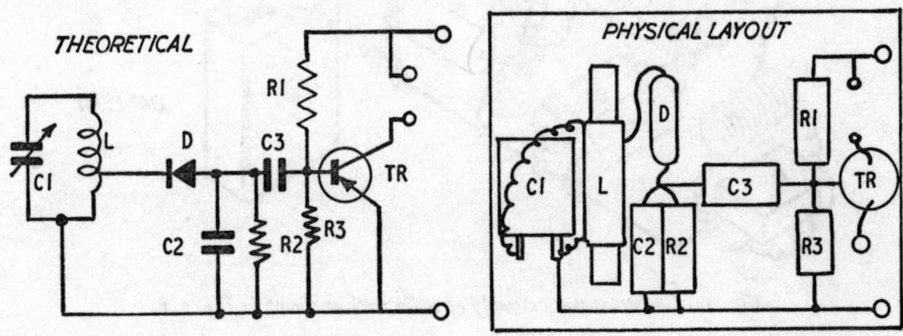

Fig. 2.1. *Comparison of theoretical circuit with physical layout of the same circuit. Note how the actual component shapes replace the symbols in the drawing on the left. The circuit shown is that of the radio receiver described in Chapter 12.*

One thing the beginner should avoid is trying to crowd components too closely together. It is much easier to plan a layout with the components spaced well apart, and also considerably easier to do a soldering job on such a layout to complete the wiring up. The ability to crowd components and produce a more compact circuit will readily come with practice.

Christmas-tree Assembly

This is a crude form of assembly which can be used for simple experimental circuits where not too many components are involved. It involves simply connecting the components together by their leads, soldering each joint, and using further individual lengths of wire, as necessary, to complete the circuit (Fig. 2.2.). All joints should be soldered—although in some cases or certain parts of the circuit, twisted connections may suffice. The result is a very messy arrangement which can only be regarded as temporary—for testing whether a particular circuit will work satisfactorily or not. It is also a dangerous arrangement since the bare wire joints can easily be touched together to cause shorts which could damage or destroy some of the components. Nevertheless

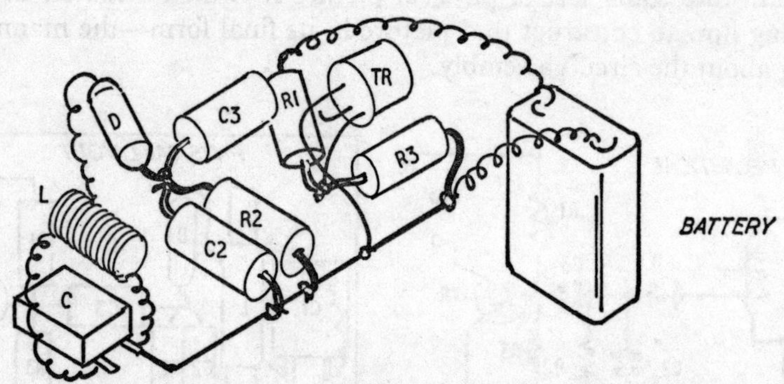

Fig. 2.2 Christmas-tree assembly of radio receiver shown in Fig. 2.1.

it is often used by experts for rapid construction of experimental or test circuits. It is not recommended for general use, even for experimental circuits, unless provision is made to insulate all bare joints by wrapping them with insulating tape. It is certainly not satisfactory for any form of permanent circuit assembly.

Tackboard Assembly

This again is a very crude form of circuit assembly, more suited for rapid experimental work than anything else. The basis of this form of construction is to draw out the physical disposition of the various components on to a piece of $\frac{1}{8}$" ply (or even $\frac{1}{4}$" thick sheet balsa) of suitable size to act as a mounting base and to mark the points at which connections have to be made to other components or to main wiring. Copper or brass nails are then driven into the ply at these points (Fig. 2.3). Where components have to be mounted to the ply panel these can be secured with screws or small bolts, as most convenient, or even in the case of small transformers, stuck down with a suitable adhesive. Where a valve base has to be mounted a hole will have to be cut in the panel to clear the tags on the bottom of the base. Small holes can then be drilled through the panel around the base to bring wiring connections up from the base tags to the face of the board for joining to the appropriate connection points of copper or brass nails.

All joints are completed by soldering to the copper nails. There is also room to wrap component leads around the nails where necessary, and the final assembly should be quite rigid and permanent. It will

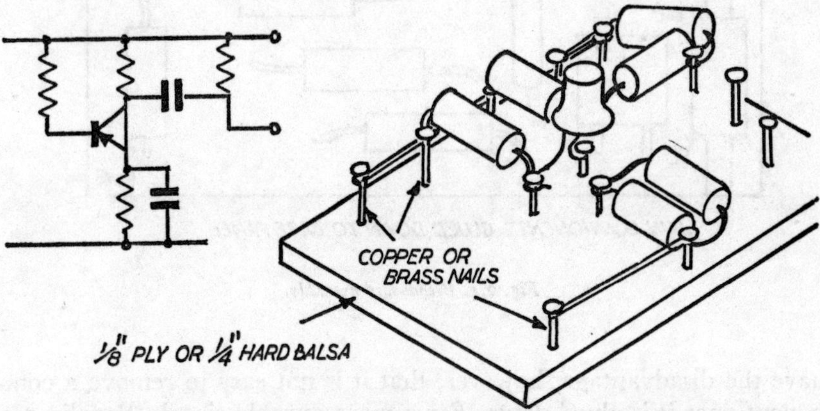

Fig. 2.3. *Tackboard assembly. The theoretical circuit is shown on the left.*

usually look rather untidy and amateurish, and the ply base will invariably become scorched around each nail due to the heat conducted down to it during soldering. The method is, however, a practical one; quick and easy, and it allows the widest possible scope for component layout and positioning. Ply itself, however, is not the best of insulators and so such an assembled circuit should never be kept or used in anything but a really dry atmosphere.

Breadboard Assembly

This is a similar form of assembly except that the base is best cut from hardboard rather than ply and all the main components themselves are fastened down to the board. Thus valve holders, coil formers, and so on are secured by bolting them to the board. Other components are then glued in place with a suitable adhesive like Bostik or Araldite. All wiring connections are completed between the components themselves, just like the Christmas-tree arrangement, except that the whole assembly is given rigidity by virtue of the fact that all the main components are rigidly mounted on the base (Fig. 2.4).

The breadboard assembly can make quite a satisfactory arrangement for small circuits where not too many components are involved. It does

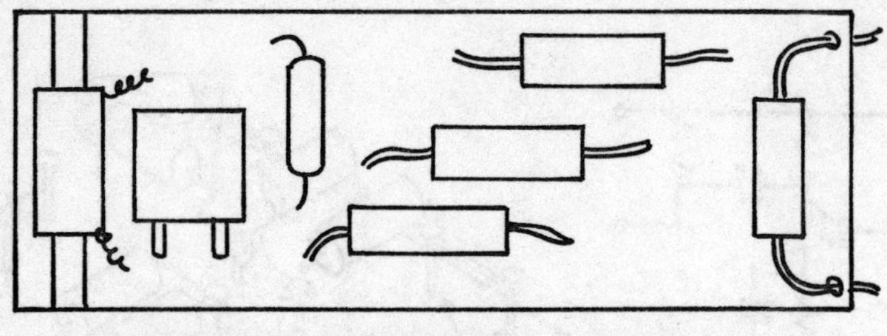

Fig. 2.4. Breadboard assembly.

have the disadvantage, however, that it is not easy to remove a component once it is glued down. For a more durable circuit, Paxolin can be used for the base panel instead of hardboard. This type of assembly is suitable for permanent circuits where the whole can be enclosed in a suitable box or case.

Pegboard System

This is a superior version of tackboard assembly which is readily suited for the construction of both experimental and permanent circuits. Its main disadvantage is that it requires a rather large area of baseboard to accommodate the components. The system in other words is not suited for compact assemblies.

The base panel is cut from standard pegboard or drilled hardboard where the holes are spaced at $\frac{3}{4}''$ intervals. The softer grade of pegboard is preferable since it is less brittle than hard pegboard. Hard pegboard is readily identified by its darker colour.

Connection and mounting points for the individual components consist of brass pillars, usually made from $\frac{3}{16}''$ diameter brass rod which has been blind drilled and tapped and cut to length.* Pillars are fitted to the baseboard either by pushing into a hole or mounting via a 6BA screw and two washers (Fig. 2.5). The latter method is used for permanent assemblies. Push-fitting of pegs is satisfactory for temporary or experimental circuits. Larger components which have to be bolted down can also be mounted on pillars so that all the wiring can be completed on the side of the board where all the components are mounted.

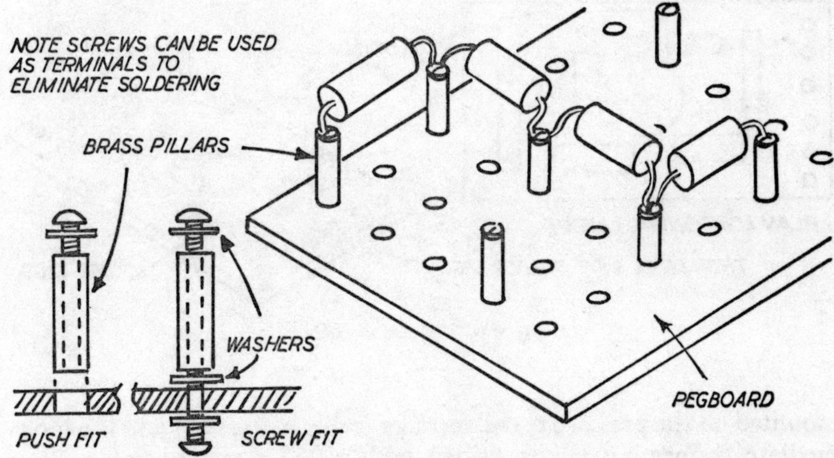

Fig. 2.5. *Pegboard assembly.*

The pegboard system is quite versatile and is adaptable to almost all types and sizes of circuits, although it tends to become unwieldy where a large number of components have to be accommodated. In such cases the overall size of panel required can be reduced by using two or even more panels in layers, one on top of the other, separated by and mounted on, longer pillars. A two-stack pegboard assembly, for example, could be based on "folding" the theoretical circuit in half to complete the physical circuit.

There are also other forms of pegboard specially made for circuit assembly, including pegboard or pre-drilled board with a printed circuit backing to which components can be mounted directly.

* Suitable pillars for pegboard assembly are made by the Weir Electrical Instrument Company, Ltd, Bradford-on-Avon, Wilts., and are also available from radio suppliers.

Tagboard Assembly

This is a simple professional type of assembly readily suitable for amateur construction where the number of components involved is not too great. The base panel is cut from $\frac{1}{16}''$ Paxolin (or $\frac{3}{32}''$ thick Paxolin for larger sizes) and the component positions marked out. Holes are then drilled or cut to accommodate such components as can be directly

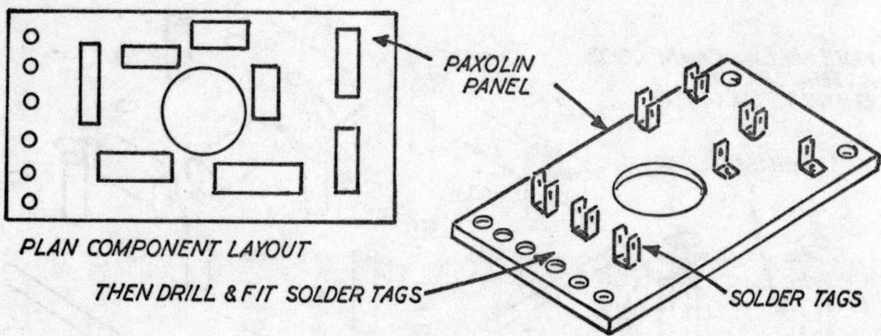

Fig. 2.6. *Tagboard assembly.*

mounted on the panel (e.g. coil formers, valve bases, etc.) and the intermediate connecting points drilled with a $\frac{3}{32}''$ diameter hole. These connecting points are completed by riveting or eyeletting a soldering tag in place, as shown in Fig. 2.6. Alternatively, the tags can be mounted by means of small nuts and bolts instead of riveting. Wiring up of the complete circuit then follows as with the previous systems.

Bus-bar Assembly

This type of assembly is particularly suited to transistor receiver assembly or similar circuits where a considerable number of components are joined to a common wire or wires (usually at the top and bottom of the circuit on the circuit diagram). A Paxolin panel is used. This is drilled for accommodating components to be mounted by screws, and also to take the common wires of the circuit in the form of bus-bars (see Fig. 2.7). These bus-bars should be cut from $\frac{1}{16}''$ diameter tinned copper wire with the ends bent at right angles to pass through holes in the Paxolin panel, and then turned over to hold the bus-bar in place. All the floating components are then soldered directly to the appropriate bus-bars as far as possible, using riveted or bolted-on solder tags for intermediate connection points, where necessary. It will often be found with bus-bar assembly that the physical circuit can be laid out almost exactly the same as the theoretical circuit.

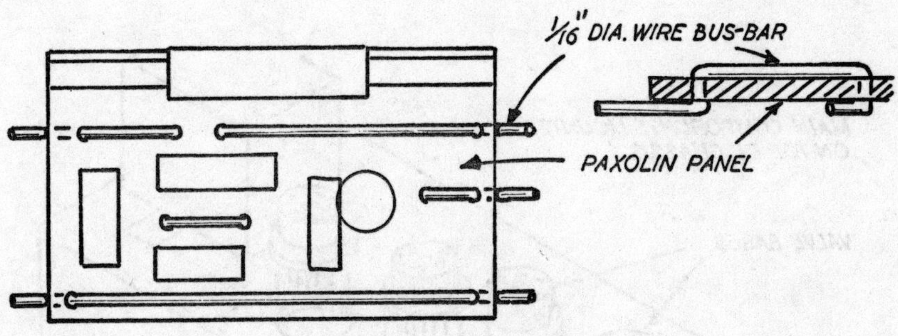

Fig. 2.7. Bus-bar assembly.

Chassis Assembly

This is the traditional method of assembly for professional valve circuits. The base panel is a metal chassis, usually aluminium, with turned down edges to give rigidity (Fig. 2.8). The chassis is drilled and cut to take mounted components, and also to accommodate intermediate solder tags, etc., which may be necessary. The floating components are then wired in place between the mounted components and the fixed solder tags. The latter also provide a common earthed connection for the circuit, unless insulated from the chassis when fitted.

This type of assembly is not greatly used for amateur work, except for permanent valve circuits, although it is still probably the best for permanent circuits of this type.

Printed Circuit Assembly

This is the modern type of circuit favoured for professional work and particularly for transistor circuits. The mounting panel is a Paxolin or glass fibre plate on which the complete wiring circuit has been duplicated in copper foil on one side. Connection points are represented by drilled holes and components are mounted by pushing their leads through the appropriate holes and soldering to the copper foil. Surplus wire is cut off. Mounting of the component in this manner automatically connects it into the circuit (Fig. 2.9).

The only real snag is that preparation of the printed circuit itself is something of a skilled job, particularly the drawing out of the required

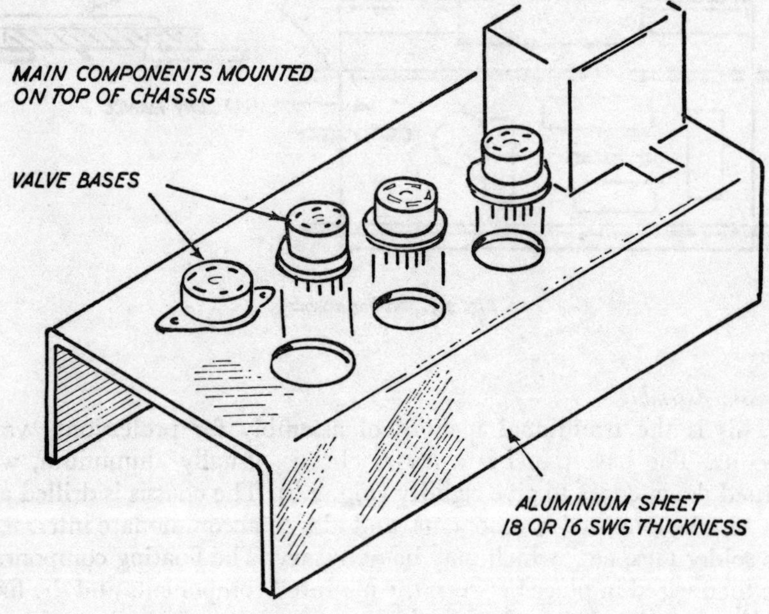

Fig. 2.8. Chassis assembly.

circuit on the panel. Printed circuit stock material consists of a laminated plastic panel fully coated on one side with copper foil. It is necessary to reproduce a drawing of the circuit wiring on this copper face and then etch away the unwanted copper to reproduce a final printed circuit. The actual making of a printed circuit is fairly straightforward; but drawing an original printed circuit to match a given theoretical circuit design is another matter.

Basically the theoretical circuit has to be laid out in the form of a physical drawing accommodating all the components and completing the connections with solid lines which do not cross (Fig. 2.10). This is not as impossible as it may seem at first sight since the components themselves act as bridges between conductors; if it is absolutely impossible to avoid crossing conductors, this connection can always be made with a separate flying lead soldered to the printed circuit.

Having prepared a suitable circuit drawing in this manner and

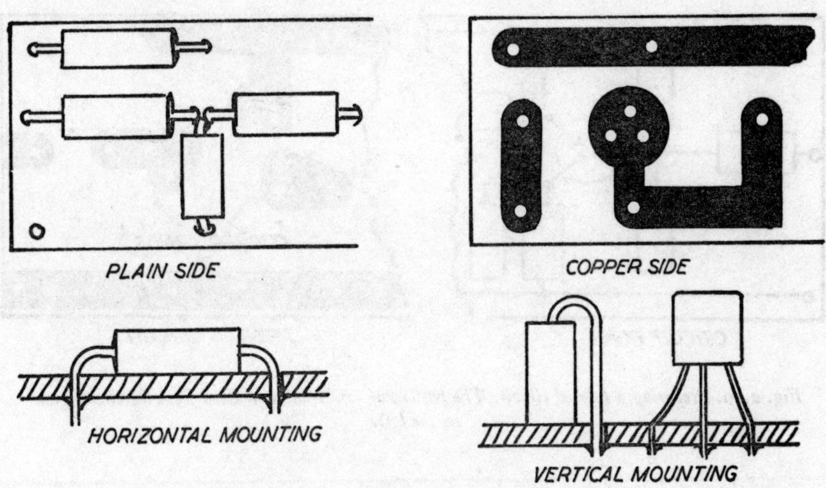

Fig. 2.9. Printed circuit assembly.

marked the position of the component leads, a mirror image is transferred to the copper side of a panel of printed circuit stock material. That is to say, the circuit as drawn on the copper is reversed left to right compared with the original circuit. This is because the wiring appears on the *back* of the panel with the components mounted on the *front*—thus the wiring up pattern is laterally inverted.

Once the circuit pattern is transferred to the copper the complete pattern of conductors is then painted in with model aircraft dope or resist ink and allowed to dry. The complete panel is then immersed in a solution of ferric chloride and left until all the exposed copper is eaten away. It can then be removed, washed and dried, and the dope or resist ink covering the circuit pattern removed with a suitable solvent.

The result is a panel on which the complete wiring circuit is reproduced in copper. It only remains to drill holes through the panel (working with the copper side up) to take the various component leads.

The serious experimenter is thoroughly recommended to use printed circuit assembly for all transistor circuits, where practicable. Starting with elementary circuits involving only a minimum of components he

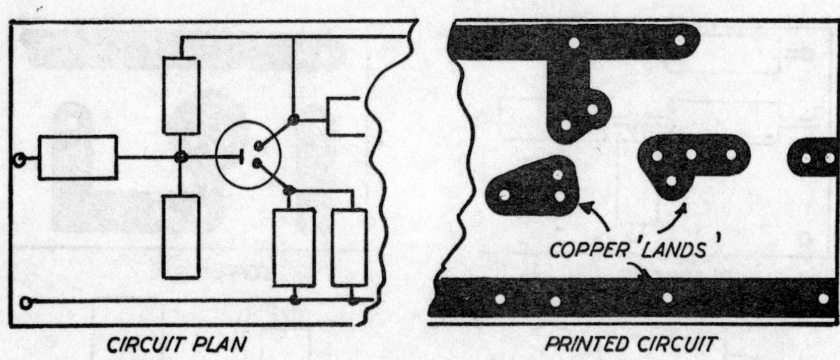

Fig. 2.10. *Preparing a printed circuit. The pattern of conductors is derived from the circuit plan on the left.*

will soon become familiar with the technique of drawing a suitable circuit, whilst preparing the final panel by etching is a straightforward process. In many cases where kits of components are available for constructing electronic devices, particularly radio receivers, these kits also include a complete printed circuit ready drilled.

Mock printed circuit

This again is one of the cruder methods of assembly where the original physical layout is based on similar lines to the above, but the mounting plate is plain Paxolin and this is merely drilled with holes to take the component leads, etc. (Fig. 2.11). Components are then mounted by cutting off the leads on the underside of the panel, leaving about $\frac{1}{4}''$ protruding, and then turning over against the panel to hold the component in place. This provides a method of mounting all the components. Wiring is then completed with separate short wires soldered between the turned over component leads.

The resulting assembly is usually messy and very liable to mistakes in wiring up, since this wiring must be completed to a mirror image drawing. Solder tags may also have to be incorporated to accommodate leads at intermediate wiring points. Some improvement is possible by combining this method with bus-bar assembly for the main wiring leads; the bus-bars in this case being mounted on the underside of the panel.

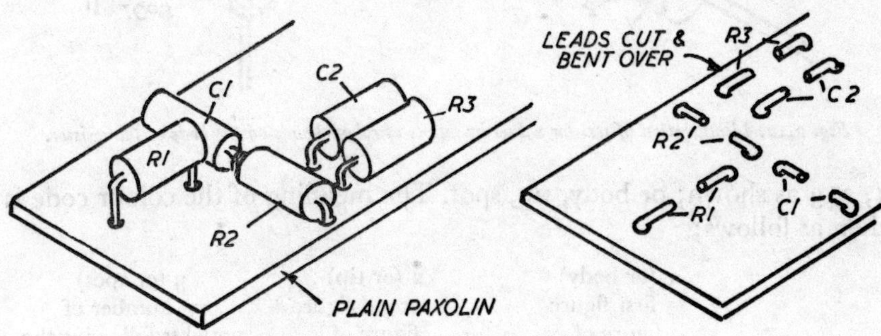

Fig. 2.11. *Mock printed circuit.*

REFERENCE NOTES ON COMPONENTS

Component values are expressed in ohms for resistors; farads for capacitors; and henries for inductances. Potential difference is expressed in volts and current in amps. Frequently, however, these basic units are too small or too large to express a particular value without having to use a lot of noughts associated with the actual figures involved. This is got round by using a prefix in front of the unit, thus:

(M) *Mega* meaning, 1,000,000 times—e.g. 1 megohm = 1,000,000 ohms.

(K) *Kilo* meaning 1,000 times—e.g. 1 kilohm = 1,000 ohms.

(m) *Milli* meaning 1/1,000th—e.g. 1 milliamp = ·001 amp.

(μ) *Micro* meaning 1/1,000,000—e.g. 1 microvolt = ·000001 volt.

(p) *Pico* meaning 1/1,000,000,000,000—e.g. 1 micofarad = ·000000000001 farad.

The value of resistors is usually marked on the body of the resistor by a colour code, either with coloured bands, or a coloured tip and different coloured spot (see Fig. 2.12). The colours are read in the order

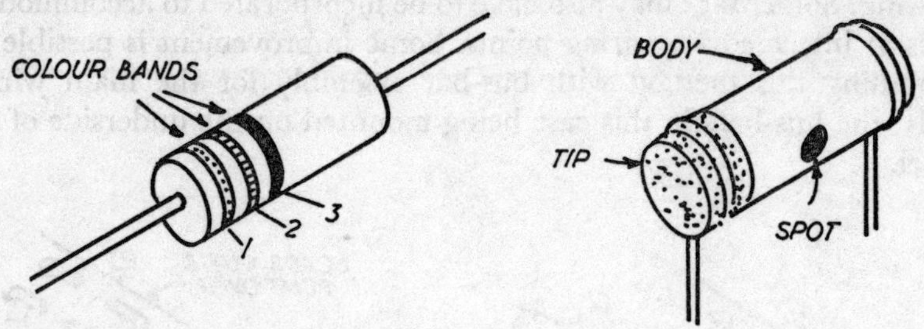

Fig. 2.12. Identification of resistor values by means of colour bands on the body of the resistor.

1, 2, 3 as shown; or body, tip, spot. The meaning of the colour code is then as follows:

	1 (or body) = first figure figure of resistance value	2 (or tip) = second figure figure of resistance value	3 (or spot) = number of noughts following the first two figures
Black	0	0	none
Brown	1	1	0
Red	2	2	00
Orange	3	3	000
Yellow	4	4	0000
Green	5	5	00000
Blue	6	6	000000
Violet	7	7	0000000
Grey	8	8	00000000
White	9	9	000000000

Example: suppose the resistor colour code reads red, violet, blue. The value is interpreted as   2   7   000000
i.e.   27,000,000 ohms or 27 megohms

In addition resistors may have a silver or gold tip or coloured band. These are not read as part of the colour code, but merely designate that the resistor is made to close tolerance—within 10% of the stated value in the case of a silver band, and within 5% of the stated value in the case of a gold band. Without such a silver or gold band the normal manufacturing tolerance for resistors is 20% of the stated value. This is a plus or minus tolerance (i.e. either side of the stated value).

Standard resistor values are produced in steps which provide an approximately constant *percentage* change in resistance rather than a simple arithmetical difference. These basic steps are:

10 12 15 18 22 27 33 39 47 56 68 82 100

Thus a range of resistors from 10 to 100 would follow values as above, From 100 to 1000 ohms the range of values would be 100, 120, 150. 180, 220, etc. From 1 kilohm standard values would be 1K, 1·2K 1·5K, 1·8K, 2·2K, 2·7K, . . . and similarly from 1 megohm up.

Values of capacitors are normally marked on the body in actual figures, except in the case of variable capacitors. Some capacitors, however, may be identified by a colour code similar to the resistor colour code.

There are various types of construction. Ceramic capacitors, which employ a small ceramic tube or disc as the dielectric with the inside and outside of the tube coated to act as plates, are restricted to small values. Another type is wound in the form of a tube comprising paper inter leaved with aluminium foil. There is also the mica type which is flat in shape and consists of silver foil interleaved with mica sheets. Both of these types are coated with a moulded or waxed casing to seal and waterproof them. For large values electrolytic capacitors have to be used, normally enclosed within a metal case. The dielectric is actually formed after assembly of the complete unit by passing a direct current through the unit to plate out a thin dielectric film. This also makes them sensitive to the direction of current flow, so electrolytic capacitors must always be connected in a circuit the right way round (the positive end or connection being marked on the case). Dielectric capacitors are also sensitive to working voltage and must not be used for voltages higher than those specified for the type.

In the case of variable capacitors the small compression or "postage stamp" types are comprised of interleaved aluminium foil and mica, and their effective capacity can be varied by adjustment of a screw bearing against the interleaved stack. On larger physical sizes air is used as the dielectric and the capacitor consists of two pivoted plates, or rather sets of plates, one fixed and one movable, and intermeshing with the fixed set, but not actually touching. The effective capacity is then determined by the relative position of the movable plates. Turning these to the maximum open position gives maximum capacity, and vice versa.

Important Notes on Soldering

The successful and consistent working of any electronic circuit depends on all wiring connections being properly soldered. Only an electric iron should be used, of a size suitable for the job. For most assemblies a 40 watt iron is about right, with a $\frac{1}{8}''$ diameter bit. For smaller assemblies, such as transistor circuits where components may be more crowded together, a 20 watt iron with a $\frac{3}{32}''$ bit may be more easy to manipulate.

It is important to use only resin-cored solder. Acid-type flux should *never* be used for electrical work as it will inevitably lead to corrosion.

A successful soldered joint depends on (i) the use of a hot iron with suitable solder (electrical quality, resin cored), and (ii) cleanliness of the leads being soldered. All leads and solder tags, should be cleaned with emery paper or a small file before attempting to solder, whether they are tinned or not, and whether they appear clean or not.

Provided (i) and (ii) above are followed, a soldered joint should be completed in about 2 to 3 seconds. If it takes longer, then either the iron is not hot enough, or the joint is not clean to start with. Excessive heat applied to a lead, caused by holding the iron in place too long, can damage components, particularly transistors. Where soldering is completed in 2 to 3 seconds no heat damage will occur, even with transistors, provided the iron is not applied to the leads closer than about $\frac{3}{8}''$ to the body. For this reason transistors should always be mounted with fairly long leads—and never flush with a printed circuit panel. A heat sink is not required when soldering transistors if the correct technique is employed.

Soldered joints which dry with a rough, uneven almost crystallized appearance are "dry" joints and quite unsatisfactory. They are usually caused by the iron not being hot enough. If solder does not stick to a joint, the reason is simply that the surfaces are not clean. A joint where the solder collects in a blob instead of flowing smoothly over the joint is also usually a sign of a dirty joint. It can also be caused by trying to apply too much solder with an iron which is not hot enough, or too small to hold enough heat to melt the amount of solder being used.

CHAPTER 3

INSTRUMENTS

THE two main instruments used for measuring DC quantities in electrical circuits are the *ammeter* and *voltmeter*. The ammeter is a series-connected instrument—it is connected into a circuit in series with other components in that circuit. The voltmeter is a parallel-connected instrument which measures a potential difference when applied across a source of potential or potential difference (Fig. 3.1).

Basically the two instruments are identical, comprising a coil of many turns of fine wire wound on to an aluminium former and pivotally mounted between the poles of a permanent magnet. Current passing through the coil generates an electromagnetic field. This reacts with the

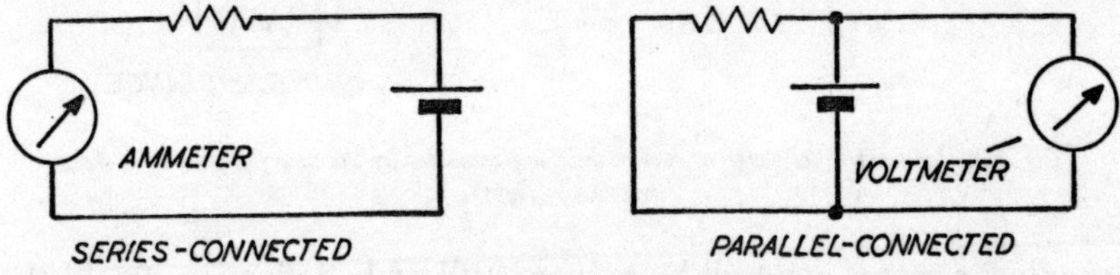

Fig. 3.1. *Circuit connection in series for an ammeter (left) and connection in parallel for a voltmeter (right).*

permanent magnet field and causes the coil to rotate against the action of a spring to a degree proportional to the amount of current flowing. A pointer attached to the coil swings with it and can indicate the coil movement on a suitable graduated scale.

Such a system obviously reads current directly, and so can be calibrated as an ammeter. Any electrical resistance offered by the coil winding must obviously affect the accuracy of current reading, and the coil resistance must therefore be kept low (5 ohms is a typical figure). Coil resistance can then be regarded as negligible in measuring current values in a practical circuit. However, the range of the instrument is also governed by the coil resistance. Thus for ranges of measurement up to about 5 milliamps the meter coil can carry the current directly. The same coil, however, would overheat or burn out if called upon to carry much higher currents, demanding either a more robust coil winding or some protection to the low resistance coil to limit the current which can flow through it.

The coil can be shunted by connecting a suitable resistor in parallel so that only a proportion of the total current passes through the meter coil. Thus in a 5 ohm coil meter with a range of 0–5 milliamps shunted with a resistor equal to the coil resistance, the actual current flowing

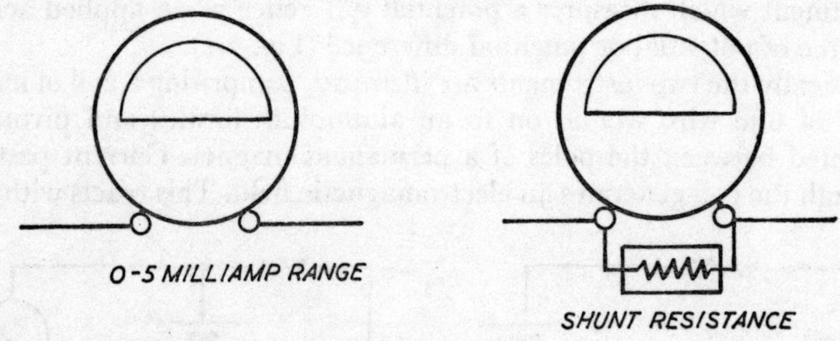

Fig. 3.2. *Increasing the range of a 0–5 milliamp ammeter by the incorporation of a shunt resistance (right).*

through the meter coil will be halved. Full scale deflection would then correspond to 10 milliamps. That is, the addition of a shunt has increased the range of the same meter movement from 0–5 milliamps to 0 to 10 milliamps. A whole range of instruments can thus be produced around a single meter movement by using different shunt resistances, which are usually incorporated inside the meter.

Equally, the range of a 0–5 milliammeter can be increased, if required, simply by connecting a suitable shunt resistance across its terminals (Fig. 3.2). The shunt resistance required can readily be calculated from the formula

$$\text{shunt resistance } R_s = \frac{R_c}{(N-1)}$$

$$\text{where } R_c = \text{coil resistance}$$

$$N = \text{the number of times the original scale range is to be multiplied.}$$

If the coil resistance is not known—it is often marked on the meter dial or face—it will have to be found by experiment. The simplest way is to find what value of resistor connected as a shunt across the terminals will exactly halve the meter reading under the same conditions. The coil resistance will then be equal to the shunt resistance used.

To work as a voltmeter, the same movement can be used, but in this case the coil has to carry a current which is proportional to the voltage being measured. In other words, the meter movement is still initiated

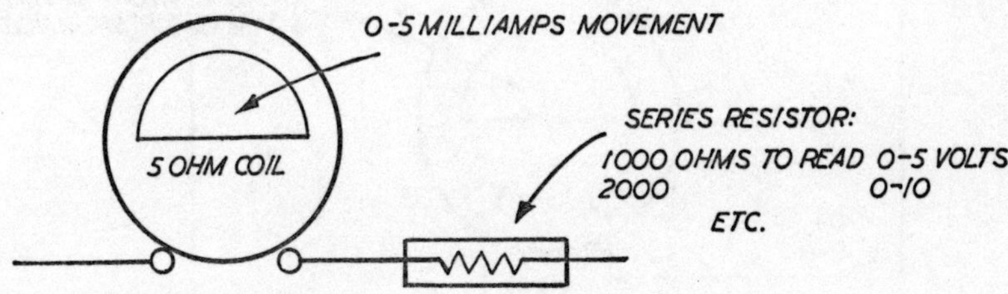

Fig. 3.3. *Conversion of milliammeter into voltmeter.*

by the value of the current flowing through it, but this current has to be related in some way to the voltage producing that current. This is done by combining a high resistance in *series* with the meter coil (Fig. 3.3). Since current flow = volts/resistance, the total resistance of the meter (series resistance plus coil resistance) can be selected to give the full scale meter current for the maximum voltage required to be measured. Thus if the full scale current is 5 milliamps the total resistance required

$$= \frac{\text{volts}}{\cdot 005}$$

Hence for a 0–5 volt meter the total resistance required would be 5/·005 = 1,000 ohms; for a 0–10 volt meter 10/·002 = 2,000 ohms; and so on.

In practice the meter resistance is virtually negligible compared with the total resistance and so such calculations can be taken as showing the actual value of the shunt resistance required. This is normally incorporated inside the meter case. On the other hand, exactly the same principle can be applied to a milliammeter to turn it into a voltmeter by connecting an external series resistance of appropriate value. Basically, therefore, given a single 0–5 milliamp ammeter one can adapt this to read other current ranges by adding shunt resistors of appropriate value; and also as a voltmeter with a variety of ranges by adding series resistors of appropriate value. This is the principle of the multi-purpose test meter, which is usually expensive to buy. It is a simple enough matter to make one from a single 0–5 milliammeter mounted in a box with external connections for different scale ranges as shown in Fig. 3.4

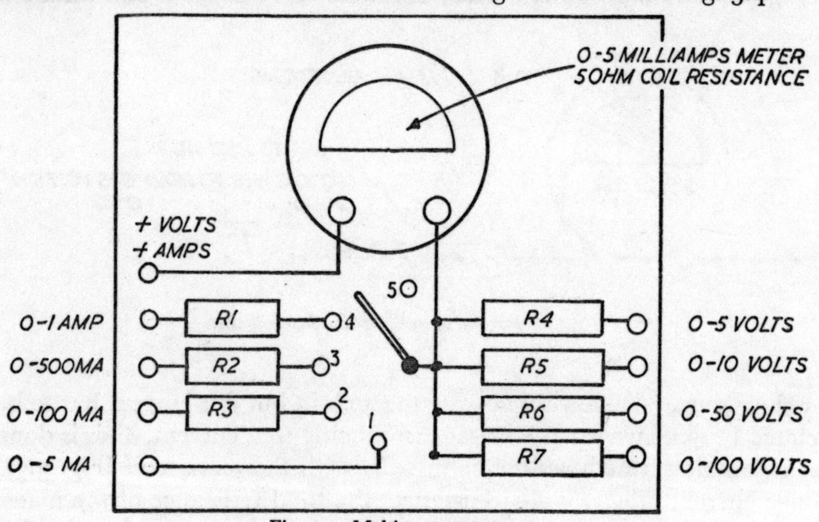

Fig. 3.4. *Multi-purpose test meter.*

Components List:
*R1 14″ length of 20 s.w.g. 56 44 copper–nickel resistance wire
*R2 9″ length of 30 s.w.g. copper wire
*R3 14¼″ length of 24 s.w.g. copper wire
 R4 1,000 ohms
 R5 2,000 ohms
 R6 10,000 ohms
 R7 20,000 ohms
      * Wound into a convenient coil form

Such meters are quite suitable for all normal DC current and voltage measurements. They are not particularly suitable for measuring very low values, however. Thus measurement of the leakage current of a transistor (see Chapter 6) will scarcely produce any appreciable movement of the pointer at all on a 0–5 milliammeter. For this a microammeter will have to be used with a much more sensitive movement which measures in millionths of an amp (microamps) rather than thousandths (milliamps). It is a more expensive instrument than an ordinary milliammeter.

The same with measuring voltages. In the case of a voltmeter the sensitivity is directly related to the resistance per volt. The ordinary voltmeter usually has a value of 1,000 ohms/volt; a 0–5 volt meter has a total resistance of 5,000 ohms, and so on. Some more expensive instruments may have a value of 3,000 or 5,000 ohms/volt and are thus more sensitive since the basic meter movement is producing a full scale deflection on much less current. Even this is still not sensitive enough to measure very low voltages, however.

To get down to very low voltages it is usually necessary to employ an electronic voltmeter, which is basically an amplifier which magnifies the tiny current obtained from the original very low voltage to a level where it will operate a conventional meter movement. Such a circuit is shown in Fig. 3.5 based on a Mullard design with an effective input resistance of one million ohms per volt. It is thus an extremely sensitive instrument capable of measuring down to *microvolts*.

The range offered by this meter is from 0 to 32 volts in eight stages set by the position of switch S_1. These ranges are:

$$
\begin{array}{rll}
\text{position } 1 & -\ 0\text{–}10 & \text{mV} \\
2 & -\ 0\text{–}32 & \text{mV} \\
3 & -\ 0\text{–}100 & \text{mV} \\
4 & -\ 0\text{–}320 & \text{mV} \\
5 & -\ 0\text{–}1 & \text{volt} \\
6 & -\ 0\text{–}3{\cdot}2 & \text{volts} \\
7 & -\ 0\text{–}10 & \text{volts} \\
8 & -\ 0\text{–}32 & \text{volts}
\end{array}
$$

After wiring it is necessary to set up and calibrate this instrument, which is done as follows:

    (i) Set switch S_1 to "bat.adj." and adjust resistor RV_{16} for full scale deflection. (Notice that "bat.adj." is position 6 on the switch.

    (ii) Set range switch to "10 mV" and switch S_1 to "Read +" (position 2 on switch). Adjust open circuit control RV_2 for zero reading.

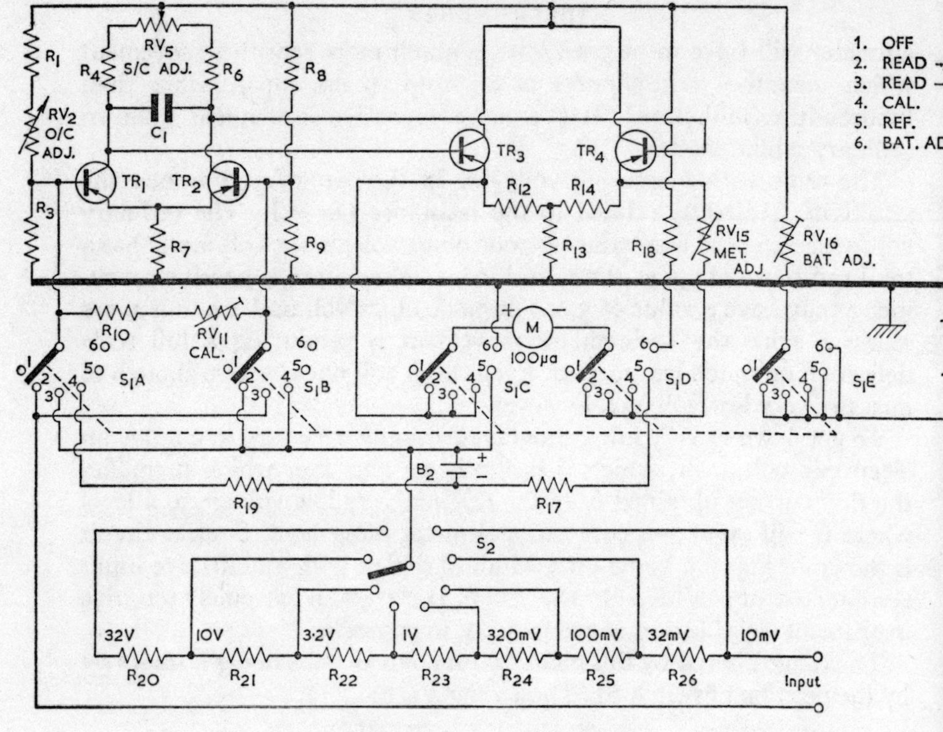

3.5. *Mullard 1 megohm per volt DC voltmeter.*

Components list:

R1	250kΩ	H.S.±1%
RV2	100kΩ	linear carbon potentiometer
R3	100kΩ	H.S.
R4	33kΩ	H.S.
RV5	25kΩ	3W linear potentiometer, W.W.
R6	33kΩ	H.S.
R7	10kΩ	H.S.
R8	300kΩ	H.S.
R9	100kΩ	H.S.
R10	10kΩ	H.S.
RV11	50kΩ	2W linear potentiometer, W.W.
R12	3·3kΩ	H.S.±1%
R13	4·7kΩ	H.S.
R14	3·3kΩ	H.S.±1%
RV15	5kΩ	3W linear potentiometer, W.W.
RV16	5kΩ	3W linear potentiometer, W.W.
R17	22kΩ	H.S.
R18	72kΩ	H.S.
R19	2·2MΩ	H.S.
R20	22MΩ	H.S.±1%
R21	6·8MΩ	H.S.±1%
R22	2·2MΩ	H.S.±1%
R23	680kΩ	H.S.±1%
R24	220kΩ	H.S.±1%
R25	68kΩ	H.S.±1%
R26	22kΩ	H.S.±1%

TR1–TR4:
Mullard BCZ 11 transistors
B1, 9 volts
B2, 1½ volts
C1   0·25μF, paper, 100 volt working (or less)

(iii) Short circuit the input terminals and adjust RV5 for zero reading.

(iv) Repeat (i) again, if necessary.

(v) Set switch S1 to "ref" (switch position) and note meter reading.

(vi) Turn switch S1 to "cal" (position 5) and adjust RV11 until the meter reads the same as in (v) above.

(vii) Connect instrument terminals to known emf (i.e. a battery of known voltage) and with the range switch (S2) at the correct position turn switch S1 to "read +" (position 2) and adjust RV15 for correct meter reading.

Once initially set up correctly in this manner the meter should need no further adjustment, except for a periodic adjustment of the open circuit control RV2.

The same basic circuit design can be used for a simpler type of assembly, eliminating the five-bank, six-position rotary switch and making provision to set up the calibration by temporary external connections.

Simple High-sensitivity Voltmeter

An alternative, and much simpler, design of high resistance voltmeter is shown in Fig. 3.6. This is, in effect, a standard bridge circuit with the meter in the centre arm. The circuit is balanced by adjusting R1 and S1. After this any variation in impedance between the emitter and collector of the transistor will upset the balance and cause the meter to be deflected. The variable resistance R2 acts as a sensitivity control for the meter. The base of the transistor draws current from the supply or input. Since the emitter to collector impedance varies with the base current, changes in input voltage produce unbalance in the bridge circuit and corresponding deflection of the meter.

The incorporation of three series resistors in the input circuit, selected by switch S2, provides three different ranges of measurement. The resistor values specified give ranges of 0–1v, 0–10v and 0–100v with corresponding input resistances of 22 to 150 kilohms per volt.

A further refinement is the incorporation of a warm/cool switch in the bridge circuit. This is used to take care of the fact that the operating conditions of the transistor is dependent on temperature, and so with this switch the instrument can be used over a wide temperature range.

The circuit is a very simple one to wire up. The meter, two potentiometers (sensitivity and zero set controls), range switch and on/off and warm/cool switches should be mounted on a panel.

To set up the instrument for use, switch on and short circuit the input terminals. Then adjust the zero set potentiometer for zero meter reading. Remove the short across input terminals and connect to a source of known potential. Adjust sensitivity control for the meter reading to correspond.

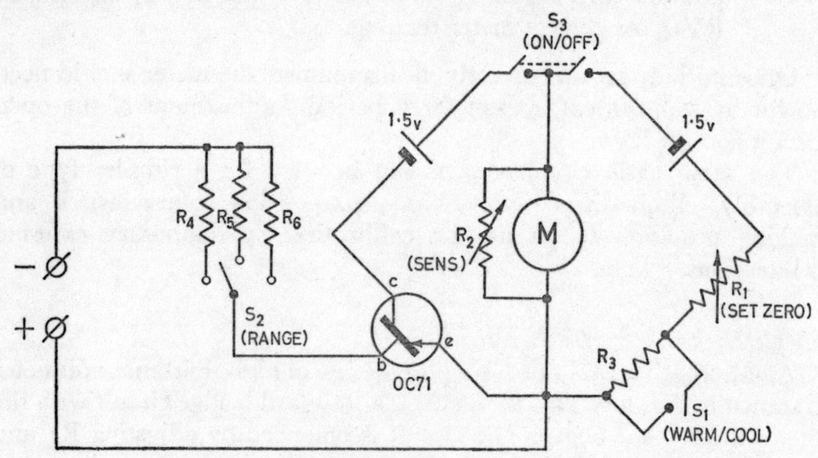

Fig. 3.6. Transistorized valve voltmeter (Mullard).

Transistor	Mullard OC71
Meter	O – 1mA d.c.
R1	50kΩ linear variable
R2	25kΩ linear variable
R3	47kΩ fixed carbon
R4	22kΩ fixed carbon
R5	220kΩ fixed carbon
R6	2·2MΩ fixed carbon
S1	Two way, single pole, toggle switch
S2	Three way, single pole, rotary switch
S3	Two way, two pole, toggle switch

Measurement of Meter Resistance

In many cases it is necessary to determine the resistance of a meter movement—for example, in order to convert it to a different range of reading via shunt or series resistors. The simplest method is to connect up the meter in parallel with a resistance box, as shown in Fig. 3.7. The 5 kilohm potentiometer is then adjusted, with the switch open, to give a full meter deflection (i.e. a full scale reading). The switch is then closed and the resistance box adjusted until the meter reading is reduced to exactly one half of its previous value. The resistance then indicated by the resistance box is equal to the meter resistance.

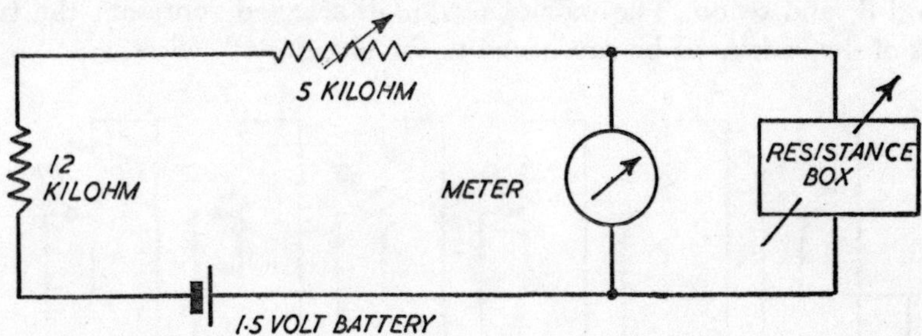

Fig. 3.7. Determining the resistance of a meter movement by means of a resistance box.

To obtain an accurate measurement of the meter coil resistance there should, of course, be no shunt or series resistors connected across or to the movement and in the case of an unknown meter it is advisable to open the case to check that no such resistors have been fitted internally. If they have, they must be disconnected for the purpose of finding the true coil resistance of the meter.

Resistance Substitution Boxes

In experimental circuits it is often highly desirable to be able to try the effect of different resistor values. This can be done by trying different resistors in turn in the circuit until one is found which produces the best results. It can be much simpler, and more accurate, to employ a resistance substitution box which, once connected, enables the resistance value inserted to be varied progressively over a wide range until an optimum value is found. This value can then be read off and the resistance box replaced by a fixed resistor of that value.

Such a resistance substitution box is shown in Fig. 3.8 and simply comprises six potentiometers mounted on a panel and wired to sockets

(Fig. 3.9). The values chosen for the potentiometers cover a range of 0 to 1,000,000 ohms (1 megohm) and enable any resistance value between these limits to be set up. Actual values can be read by calibrating the potentiometer control movement. Pairs of terminals for connection are selected according to the likely resistor value required. Thus for working with resistance values between 0 and 10 ohms, terminals 1 and 2 would be used; for working between 0 and 1,000, terminals 1 and 6; for working between 1,000 and 10,000, terminals 7 and 8, and so on. The pair of terminals selected represent the two ends of the resistor to be connected to the circuit with wires.

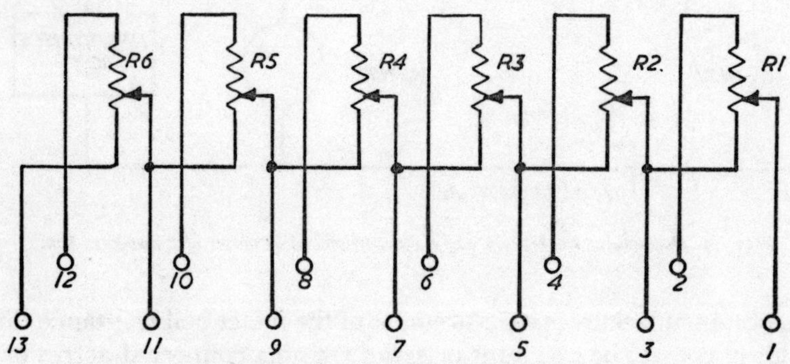

Fig. 3.8. Resistance substitution box.

Components list:
R1   10Ω, 12W wire wound potentiometer (Colvern Ltd.)
R2   100Ω, 5W wire potentiometer (Colvern Ltd.)
R3   1kΩ, 3W
R4   10kΩ, 3W
R5   100kΩ, 3W
R6   1MΩ, 1W carbon potentiometer

The only limitation with a substitution box of this type is that the current rating of the various potentiometers must not be exceeded. For the components specified this means that the voltage applied across any one potentiometer should not exceed the following values:

R1 0–10 ohms   potentiometer,   11 volts
R2 0–100 ohms   potentiometer,   22 volts
R3 0–1,000 ohms   potentiometer,   54 volts

R4 0–10,000 ohms potentiometer, 160 volts
R5 0–100,000 ohms potentiometer, 550 volts
R6 0–1 megohm   potentiometer, 600 volts

Capacitance Substitution Box

This serves the same purpose as a resistance substitution box, only this time a range of capacitance values are made available. It is not practical to use variable capacitors in a similar connection so fixed values are employed, with values selected as covering the widest range

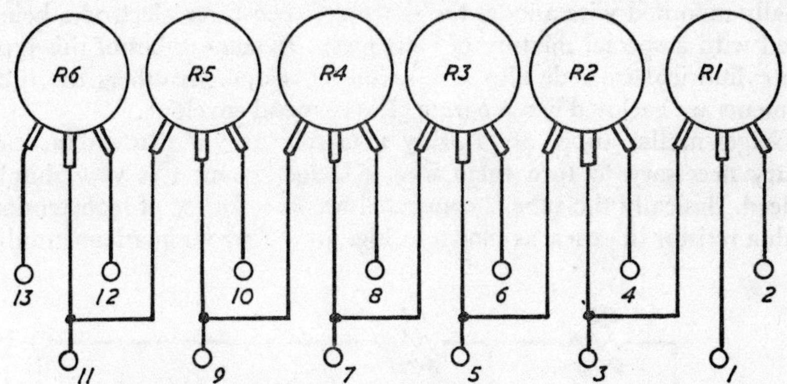

Fig. 3.9. *Wiring of potentiometers in resistance substitution box (Fig. 3.8).*

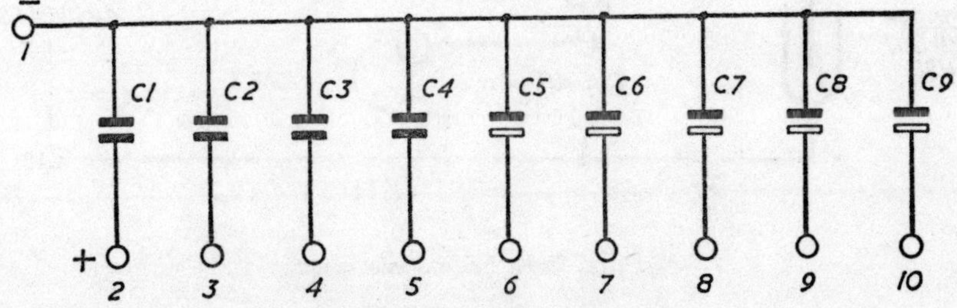

Fig. 3.10. *Connection of capacitors in capacitance substitution box.*

Components list:

C1	100pF 500 volts wkg.	C6	10μF electrolytic 50 volts wkg.
C2	0·001μF 500 volts wkg.	C7	100μF electrolytic 50 volts wkg.
C3	0·01μF 500 volts wkg.	C8	1,000μF electrolytic 50 volts wkg.
C4	0·1μF 400 volts wkg.	C9	10,000μF electrolytic 25 volts wkg.,
C5	1μF 400 volts wkg.		Mullard Type C432 series.

of requirements likely to be encountered. To simplify connection to the test circuit the output is provided by a common line to all the capacitors and the other end of the appropriate capacitor selected by plugging into the appropriate socket, as shown in (Fig. 3.10). Capacitance values available range from 100 pf to 10,000 mf in steps of 100 pf.

A Simple Geiger Counter

The heart of a Geiger counter is a special valve, known as a geiger-müller tube, which comprises a metal cylindrical cathode and an axially mounted wire anode, the space between these electrodes being filled with a special mixture of inert gases. In some tubes of this type the cylindrical cathode also acts as the envelope. In others the tube elements are enclosed in a separate glass or metal envelope.

Geiger-müller tubes are readily available and the attendant circuitry necessary to turn them into a Geiger counter is very simple indeed. Basically the tube is connected across a source of high tension with a resistor in series, as shown in Fig. 3.11. Any fluctuations in tube

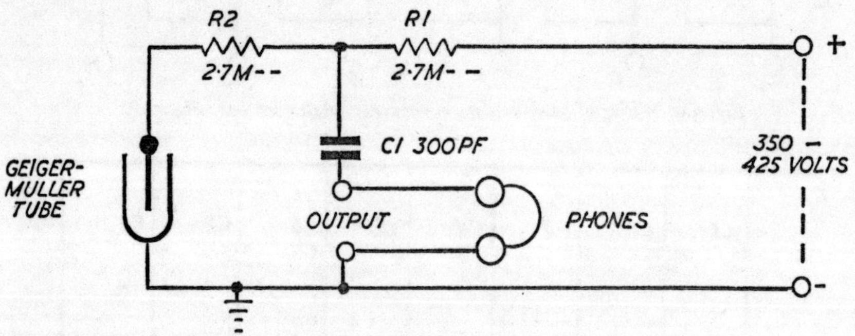

Fig. 3.11. Simple Geiger counter circuit.

current due to ionization of the gas produces an additional potential difference across this resistor, known as the "counting voltage". The addition of a second resistor and tapping off the output between the two both increases the safety factor of the circuit and produces a larger counting voltage. A capacitor in the output circuit is used to act as a DC block so that the voltage at the output terminals is that resulting from ionization only. If a pair of high impedance earphones are connected to the output terminals the "count" will be heard as a series of loud clicks.

The main limitation of the Geiger counter circuit is that geiger-müller tubes need a high working voltage—425 volts is a typical figure—which can be supplied by a suitable transformer. The complete unit must, therefore, be carefully built and enclosed within a suitable case so that there is no possibility of accidentally contacting the high-voltage connections. The counting voltage appearing at the output terminals is quite low and is of the order of 10 to 12 volts.

Geiger Counter Circuit

The Geiger counter described is really a probe unit only, with audible indication. If it is combined with a ratemeter the "count" can be presented in the form of a meter deflection. A suitable circuit for this purpose is shown in Fig. 3.12. This is capable of accommodating a counting rate of up to 30,000 counts per minute.

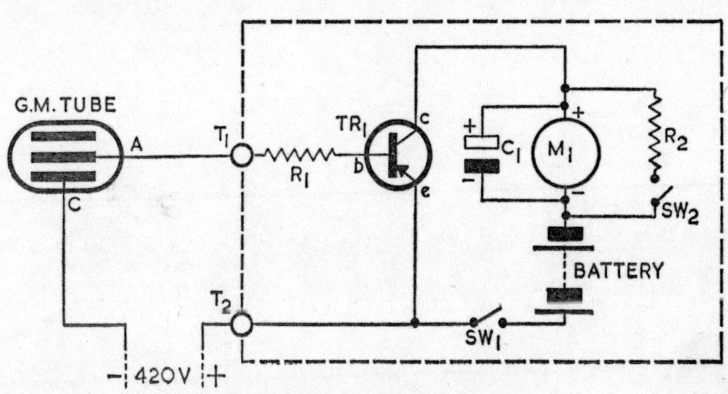

Fig. 3.12. Gieger probe ratemeter (Mullard)

Components list:

R1	$3 \cdot 3M\Omega$ ¼W
R2	250Ω
C1	$100\mu F$, electrolytic 6V wkg
M1	$50\mu A$ moving coil
TR1	Mullard OC202 silicon transistor
SW1	2 way, single pole toggle switch
SW2	2 way, single pole rotary switch
T1, T2	Screwdown terminals, one black, one red
Battery	$4 \cdot 5V$ bell battery

The circuit is a very simple one, and is basically a current amplifier. It is connected in series with the high tension supply and the geiger-müller tube itself with a load resistor R_1 to prevent the tube passing excessive currents to the transistor. The meter range is established by the shunt resistor R_2 (giving two different ranges) whilst capacitor C_1 across the meter smoothes out the current pulses through the meter and thus provides a steady reading.

The recommended transistor is a silicon type since these have a much lower leakage current than germanium transistors and thus provide more accurate meter readings over a wide range of temperatures.

The ratemeter can only be calibrated by practical means, e.g. by direct measurement against a known source of radiation. If this is available then a complete scale calibration is possible by applying the inverse square law of radiation intensity and distance. In the absence of any known source for calibration the scale readings are merely read as comparative figures.

CHAPTER 4

ELECTROMAGNETIC COILS AND RELAYS

A LENGTH of wire capable of carrying an electric current wound in the form of a coil has a wide application in electrical and electronic devices. Such a coil may be a very simple one, consisting of only a few turns of wire to provide a specified *inductance*. These coils can then form one of the two basic components of a tuned circuit for radio transmitters and receivers (see Chapter 13). Similarly, a larger number of turns of much thinner wire wound on a suitable former can provide impedance values suitable for acting as a choke in radio circuits.

In such applications coils are employed as specific components in an electronic circuit, the important value as regards the working of the circuit being the coil inductance (or impedance in the case of chokes). This is something which is difficult to calculate in designing a coil and so suitable inductance coil sizes and windings are usually determined along practical lines.* Where circuits described elsewhere in this book employ coils as inductances, the appropriate sizes of winding data are given instead of component values, as is the case with other circuit components.

Wound coils are also the basis of transformer construction, only here a large number of turns are usually employed on a pair of associated coils in order to fully saturate a common iron core with magnetic flux. This provides a magnetic, or rather inductive, coupling between the two coils thus giving a transfer of voltage and current between two parts of a circuit which are not directly connected. Transformer coupling is

---

* There is an approximate formula which can be used for calculating the size required from a single-layer air-cored coil to produce a specific inductance. This is:

$$\text{Inductance (microhenries)} = \frac{0.2\ N^2 D^2}{3.5 D + 8 L}$$

where N = number of turns in the coil

     D = diameter of coil in inches

     L = length of coil in inches

It will be noted that the actual wire diameter does not enter into this formula. This is chosen to give a suitable stiffness to the coil—for example 18 s.w.g. or 16 s.w.g. wire should be used for a coil of diameter $\frac{1}{2}''$ or greater.

used in many electronic circuits, whilst transformers themselves are also a very important subject on their own (see Chapter 9). Quench coils are also simple transformers with a 1:1 coupling ratio, used in super-regenerative detector circuits for simple radio control receivers (see Chapter 13).

The single wound coil also has numerous uses on its own, because of its ability to provide a mechanical output force from an electrical input. In other words, it is a form of transducer which can convert electrical energy into mechanical energy. Typical examples are the electro-magnet and the relay. Both are usually regarded as electrical rather than electronic devices, but since they are often associated with electronic circuits as components they qualify for description in a book on electronics. This is particularly true in the case of the relay which is commonly employed to convert a current change produced in an electronic circuit into a mechanical switching action controlling another circuit, or even the original circuit.

Coils of this type are normally wound on rigid bobbins which, besides providing a neat and compact coil, also gives maximum mechanical strength and protection against vibration and shock. Common bobbin materials are Tufnol, Paxolin, moulded Bakelite, etc., the bobbin being turned from solid stock or fabricated from sheet materials. Equally, simpler materials like thin ply can be used for the cheeks or end plates of bobbins, with Paxolin tube or even hardwood dowel for the centre or core. The main thing to remember with a built-up bobbin is that the application of the winding may impose a considerable crushing load on the core and a bursting load on the end cheeks.

Simple coils can be wound directly by piling the turns up on each other, layer by layer, as winding proceeds. Where a large number of turns are involved hand winding can be a tedious process and simple mechanical winding is usually best, as well as making it easier to produce a neater and more compact coil. The only apparatus required for machine-winding coils in thin wire is an ordinary hand drill which can be gripped in a vice. The coil bobbin is then mounted on a suitable rod, e.g. a length of studding with nuts tightened up against each cheek to lock the bobbin on to the studding. The studding is gripped in the jaws of the drill chuck. The bobbin is then rotated by turning the drill and the wire fed neatly on to the revolving bobbin. The wire used for winding must of course, be insulated. Enamelled wire is satisfactory for most purposes, but enamelled and double silk covered wire is better in the case of finer wires.

With larger coils, or better quality coils, it is usual to provide additional insulation between adjacent layers. That is, after completing a single row of winding this is covered with a layer of insulating material before the next layer is wound on, and so on. Ordinary unvarnished paper is suitable for interlays and need only be quite thin (about 1 thousandth of an inch). In the case of two or more coils wound on the same former or bobbin—where the second coil is wound on top of the first—additional insulation is usually provided between the two coils with a wrapping of PVC insulating tape.

The electromagnetic performance of such a coil is directly related to the product of the number of turns in the coil and the current flowing through the wire, or ampere-turns as this is referred to. The other important factor is the actual electrical resistance of the coil which will govern the current for any specified voltage applied to the coil. The coil resistance, in turn, is related to the size or diameter of wire used and the actual length of wire. Once all these factors have been sorted out the physical size of the coil can be calculated.

The basic relationship between ampere-turns, applied voltage and coil resistance is:

$$\text{ampere-turns} = \frac{\text{applied voltage} \times \text{no. of turns}}{\text{coil resistance (ohms)}}$$

A typical coil for a sensitive relay used in a radio control receiver circuit with a high tension voltage of 30v or more has a resistance of 4–5,000 ohms. To produce this resistance in a small coil size means using very thin wire—48 s.w.g. or a little less than two thousandths of an inch in diameter. Even so, the *length* of wire to give this resistance will still be several hundred yards, and all this length has to be accommodated on a relatively small bobbin. In practice it is usual to determine a suitable bobbin size and then fill with windings—a typical size being shown in Fig. 4.1. The actual number of turns involved will be something like 50 to 100 thousand. The relay is connected into the HT lead of the receiver circuit and so receives the full HT voltage. The ampere turns will therefore be of the order of $50,000 \times \cdot005 = 250$; i.e. the maximum current flowing through the relay coil will typically be of the order of 3 to 5 milliamps. Thus for this order of current, which is typical for sensitive relay operation, an ampere turns figure of about 250 to 500 is required.

Any change in circuit application may affect the performance. Thus if the same relay is used with, say, only 10 volts available, the ampere-turns figure will drop because the lower applied voltage will not produce anything like the same current as before with that particular coil resistance. The only way to get round this is to increase the number of turns still more (which will also increase the coil resistance by calling for more length of wire); or reduce the coil resistance by using larger diameter wire (which will increase the size of the coil for the same number of turns, but also increase the current and ampere turns).

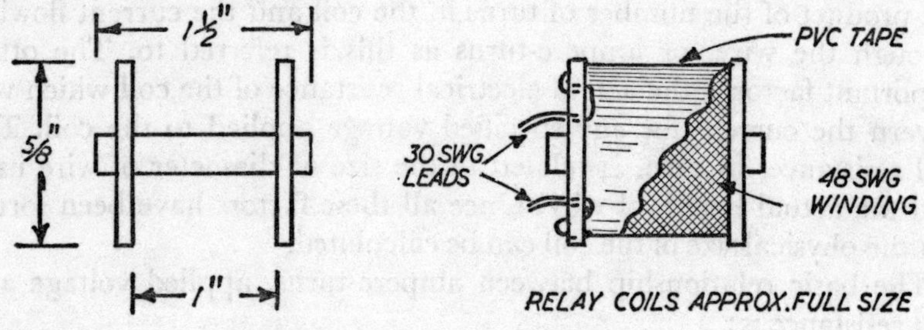

Fig. 4.1. *Typical coil for sensitive relay.*

The correct matching of a relay design to the type of circuit involved is thus very important. In practice it is usually the *current* flowing through the relay which is the critical factor since this is critically dependent on the applied voltage available and the coil resistance of the relay. The relay coil resistance, therefore, must be chosen with regard to the voltage available. It is then a matter of relay design to ensure that the relay will have sufficient ampere turns to operate properly at this current level. Thus most sensitive relays are designed to pull in properly on currents of the order of 1 or 2 milliamps. Other relays may demand more current to pull in properly, and can thus only be used in circuits where such higher current values are available.

It is seldom worthwhile making relays, particularly sensitive relays, since the design requirements are quite critical; winding the coils is a tedious process, even with a mechanical winder. When required, therefore, a relay is usually bought as a finished component, with its suitability primarily dependent on its coil resistance and operating current. The most important thing then is to adjust the relay to operate correctly and efficiently at the desired current level.

A typically simple sensitive relay is shown in Fig. 4.2. The coil is mounted on a soft iron core (often a steel bolt) attached to a soft iron bracket which carries a pivoted armature. Armature movement up and down is limited by contacts, and a light spring is fitted to tension the armature away from the face of the coil. The armature thus bears against the uppermost contact, or NC (normally closed) contact, as it is called.

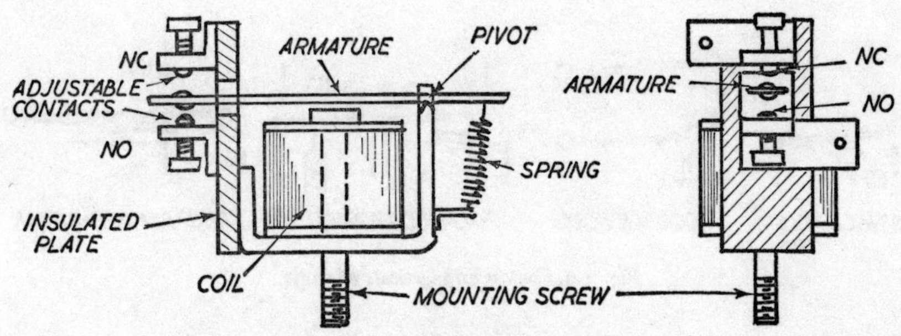

Fig. 4.2. *Typical simple relay assembly.*

When the coil is energized by the passage of a suitable current the armature is pulled downwards until it reaches the bottom of NO (normally open) contact. Thus an external switching circuit is provided through the armature and the NC and NO contacts—a circuit which opens when the relay pulls in if connected to the armature and NC contact. The circuit closes when the relay pulls in, if connected to the armature and NO contact.

The current value at which the relay pulls in can be adjusted by movement of the NC contact, which arrests the outward movement oi the armature. The closer the armature is to the face of the coil to start with, the lower the current at which it will pull in. Pull-in point is also affected by spring tension. The weaker the spring tension the lower the current at which the armature will pull in, and vice versa.

When the armature is pulled in, the closer it is to the coil face (governed by the adjustment of the NO contact) and the less readily will it drop out again when the current falls off. Again, too, spring tension will have an effect on drop out.

The object in relay adjustment is to set up the contact position so that the relay pulls in at a current substantially below the maximum current available since this will give a margin of excess current to hold the relay in firmly. Basically this is done by adjusting the top or NC contact position. At the same time, for smart response the relay wants to drop out when the current falls to a little below the pull in value—the difference between pull in and drop out current being known as the differential. In other words, adjustment should be made to give a small

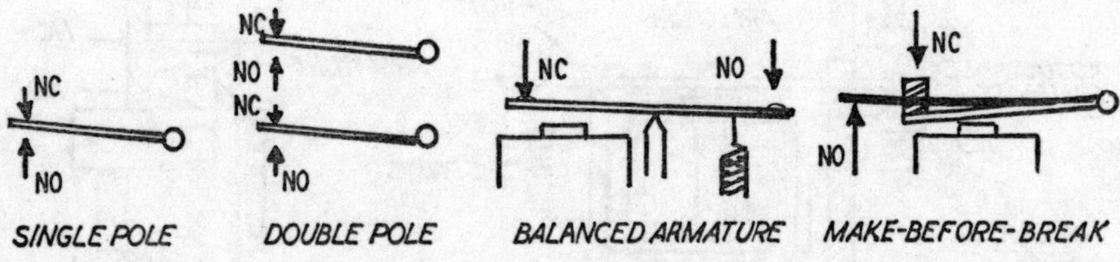

Fig. 4.3. *Contact arrangements of relays.*

differential for crisp relay response. Basically this is done by adjusting the position of the NO contact to establish a nominal drop out point, followed by adjustment of both contacts simultaneously, maintaining the same distance between them to establish a suitable differential. It may sound tricky—and it is—but the technique is soon mastered with practice. Once set up correctly the relay should then continue to function satisfactorily without further adjustment. Adjustment is critical in some circuits where only small current changes are involved, but less critical in others where the current change may be quite large. In some cases the original setting of the relay as purchased may be quite satisfactory and need no adjustment at all.

The relay type shown in Fig. 4.2 is a very simple one with a single pair of contacts. There are many variations possible by incorporating additional and separate contact sets, all controlled by the movement of the armature, although where the armature has to provide more mechanical effort for contact closure, a rather more robust design of relay is normally employed. Some typical alternative contact arrangements are shown in Fig. 4.3. There are many others, but those illustrated are the ones normally employed in simpler circuits. It should be noted that separate contact sets are insulated from one another and thus can perform as quite separate switches in different circuits.

Some design formulas which may be useful for anyone wishing to wind their own coils are summarized below (see also Fig. 4.4 for related bobbin dimensions).

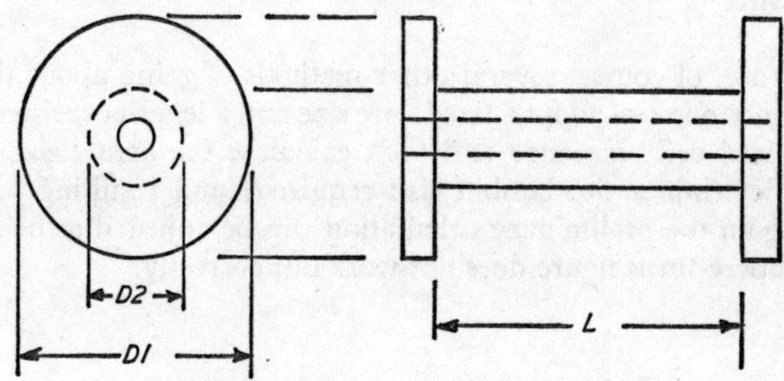

Fig. 4.4. Essential bobbin dimensions required for coil design.
(All dimensions designated in inches)

Basic Procedure in Relay-type Coil Design
  (i) "Guesstimate" or choose a wire size which should be suitable by comparing with a similar existing coil.
  (ii) Calculate the maximum number of turns which can be accommodated in the available winding space or bobbin size from the formula:

$$\text{total turns (N)} = \frac{\frac{1}{2} L (D_1 - D_2)}{d_c^2}$$

(iii) Calculate the length of wire involved from the formula

$$\text{length of wire} = \frac{1}{2} \pi N (D_1 + D_2) \text{ or } \frac{\pi L (\text{inches } D_1^2 - D_2^2)}{4 d_c^2}$$

where d_c = actual diameter of covered (insulated) wire

(iv) Find the resistance of this wire by reference to wire tables.

(v) Calculate the ampere turns from the formula

$$\text{ampere-turns} = \frac{\text{applied volts} \times \text{no. of turns}}{\text{resistance}}$$

(vi) If this is satisfactory, proceed to wind the coil. If the ampere-turns figure is too low, start the calculations again with a slightly larger diameter wire. If the ampere turns are too high, start the calculations again with a slightly smaller diameter of wire.

There are, of course, several other methods of going about the job. For instance one can adopt a fixed wire size and a length consistent with the required coil resistance and then calculate the actual size of coil (which determines the bobbin size required) and resulting ampere-turns. Again the preliminary calculation can be adjusted as necessary, if the ampere-turns figure does not work out correctly.

Table 1
USEFUL DATA FOR COIL DESIGN
Approx. turns per sq. in. winding area

s.w.g.	Enamel	Single Silk Enamel	Double Silk	Double Cotton	Amp. Rating* at 1,000 amps/sq. in.)
10	55	53	54	49	12·8
11	66	64	65	58	10·5
12	83	78	81	70	8·15
13	105	99	103	88	6·65
14	138	131	136	112	5·03
15	172	162	170	140	4·07
16	219	201	216	170	3·28
17	286	256	277	222	2·46
18	388	359	384	286	1·81
19	550	500	540	354	1·26
20	670	610	655	450	1·02
21	850	755	815	565	·80
22	1,110	975	1,040	730	·615
23	1,500	1,300	1,369	860	·45
24	1,500	1,560	1,600	975	·38
25	2,150	1,890	1,900	1,110	·31
26	2,650	2,300	2,350	1,370	·25
27	3,180	2,740	2,800	1,550	·21
28	3,900	3,300	3,330	1,760	·17
29	4,550	3,800	3,850	1,960	·145
30	5,550	4,500	4,500	2,400	·12
31	6,300	5,100	5,020	2,600	·105
32	7,300	5,810	5,625	2,830	·092
33	8,400	6,600	6,400	3,080	·078
34	10,000	7,700	7,300	3,450	·066
35	11,000	8,900	8,400	3,700	·055
36	14,500	10,400	9,790	4,100	·045
37	18,200	12,600	11,550	5,200	·036
38	23,000	15,620	13,800	5,900	·028
39	30,700	20,400	16,850	6,700	·021
40	35,500	23,000	18,750	—	·018
41	43,400	26,800	23,000	—	·015
42	50,500	30,500	26,600	—	·013
43	69,500	36,900	29,000	—	·010
44	81,500	43,200	34,200	—	·008
45	109,000	51,500	40,000	—	·006
46	192,000	65,500	47,000	—	·0045

* This column gives the maximum current for the size of wire specified, consistent with a "safe" rating of 1,000 amps per sq. ft. of conductor cross section, usually adopted for coils which are energized continuously. A higher rating can be used for coils which are only used intermittently.

Table 2

Bare Wire Data (Resistance Values)

s.w.g.	Wire dia. inches	Cross Sectional Area (sq. in.)	Resistance (ohms) per 1,000 yards	Weight (lbs.) per 1,000 yards
14	·080	·005027	4·776	58·13
15	·072	·004072	5·897	47·09
16	·069	·003217	7·463	37·20
17	·056	·002468	9·747	28·48
18	·048	·001810	18·267	20·93
19	·040	·001257	19·105	14·53
20	·036	·001018	23·59	11·77
21	·032	·000504	29·85	9·30
22	·028	·000616	38·99	7·12
23	·024	·000452	53·07	5·23
24	·022	·000380	63·16	4·396
25	·020	·000314	76·42	3·633
26	·018	·000255	94·35	2·943
27	·0169	·000211	113·65	2·443
28	·0148	·000172	139·55	1·9895
29	·0136	·000145	165·27	1·68
30	·0124	·000121	198·8	1·3966
31	·0116	·000106	227·2	1·2222
32	·0108	·000092	262·1	1·0594
33	·0100	·000079	305·7	·9083
34	·0092	·000066	361·2	·7688
35	·0084	·000055	433·2	·6409
36	·0076	·000045	529·2	·5246
37	·0068	·000036	661·1	·4200
38	·0060	·000029	849·1	·3270
39	·0052	·000021	1130·5	·2456
40	·0048	·000018	1326·7	·2093
41	·0044	·000015	1578·9	·1759
42	·0040	·0000126	1910·5	·1453
43	·0036	·0000102	2359	·1177
44	·0032	·0000080	2985	·0930
45	·0028	·0000062	3890	·0712
46	·0024	·0000045	5307	·0523
47	·0020	·0000031	7642	·0363
48	·0016	·0000020	11941	·0233
49	·0012	·0000011	21228	·0131
50	·0010	·000000785	30568	·0091

CHAPTER 5

VALVES AND SIMPLE VALVE CIRCUITS

BASICALLY an electronic valve consists of two or more elements or electrodes enclosed within a glass envelope which is evacuated to a high vacuum—rather like an electric light bulb, in fact. However, in the case of an electronic valve the elements are not directly connected to one another, and the working of the valve as an element in an electrical circuit depends on thermionic conduction (unlike a light bulb where electricity is conducted directly through a continuous filament, causing it to heat up and give out light). The proper name for a valve of this type is, in fact, "thermionic valve", although the Americans call them "tubes" or "vacuum tubes". And the basic characteristic of thermionic conduction, which gives valves such useful working properties as circuit elements, is that conduction of electricity is confined to one direction only. The device is thus a true valve in that it allows the flow of electricity one way, but shuts off the flow in the other direction. Similarly, by making the valve more complex it can regulate or control flow.

The simplest type of valve consists of just two elements enclosed within an evacuated envelope and is called a *diode*. The elements are a cathode (negative electrode) and anode (positive electrode) (Fig. 5.1). Again American terminology differs slightly. They call the anode the "plate".

The connections are peculiar to all types of devices working by thermionic conduction in that the cathode must be heated to give off electrons. Thus the cathode is made in the form of a simple heater element and is often called the filament. It is rather like a low power light bulb element in fact and is connected to its own battery or source of electricity, commonly known as the low tension supply (or A battery in American terminology). As a working component in a complete circuit the anode has to be supplied with a separate higher voltage or high tension supply (B battery), connected as shown in Fig. 5.2. It is this anode circuit, completed by electricity flowing through the valve, which comprises the working circuit of which the valve is a component.

Although both work in the same manner, there is a distinction between a battery valve and a mains valve. In the former case the filament is heated directly by the LT battery connected to it. In the case of

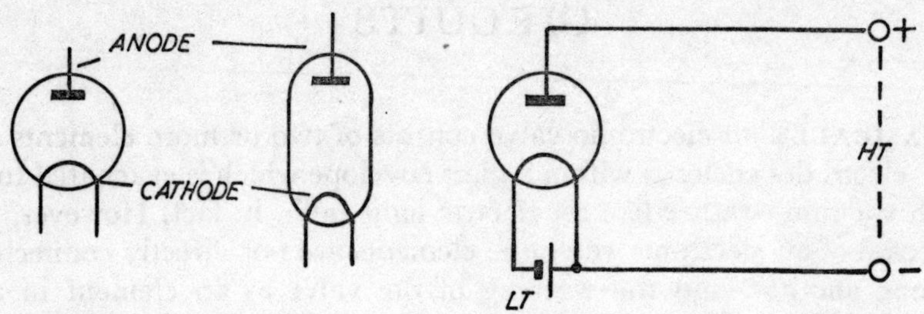

Fig. 5.1. Elements of a diode (left). The cathode is connected to low tension supply (right).

a mains valve the filament is heated indirectly by a separate coil or heater which will perform equally well on AC as well as DC. It can thus be powered from a transformer winding without rectification. The only difference as regards actual *working* is that the mains valve with its indirectly heated filament takes longer to warm up. This is noticeable on a mains receiver, for example, where all the valve filaments are

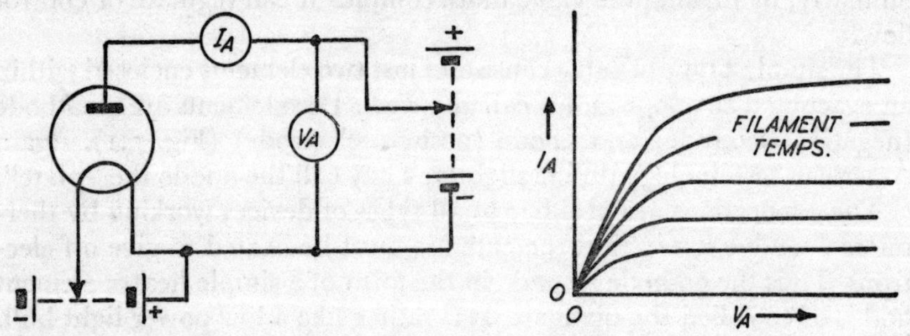

Fig. 5.2. Diode in circuit (left), showing high tension supply to anode. Performance curve on right.

indirectly heated. A battery radio with directly heated valves, on the other hand, works immediately it is switched on.

With DC working—where batteries are used for the low tension and high tension supply—the diode has the characteristic that as the anode voltage (HT battery) is increased the current flowing through the valve (known as the anode current) increases roughly in proportion, up to a point where the valve is "saturated". At this point all the electrons emitted by the cathode are being attracted to the anode, and a further increase in anode voltage can cause no further increase in anode current. The low tension (filament) voltage is fixed and simply matched to the design of the valve to ensure an adequate cathode temperature. Thus the conduction or anode current realized is a function of the applied anode voltage.

If an *alternating* current is applied to the anode the effect is somewhat different since the valve can only conduct in one direction. Thus the valve will pass positive half cycles of AC applied to the anode, since

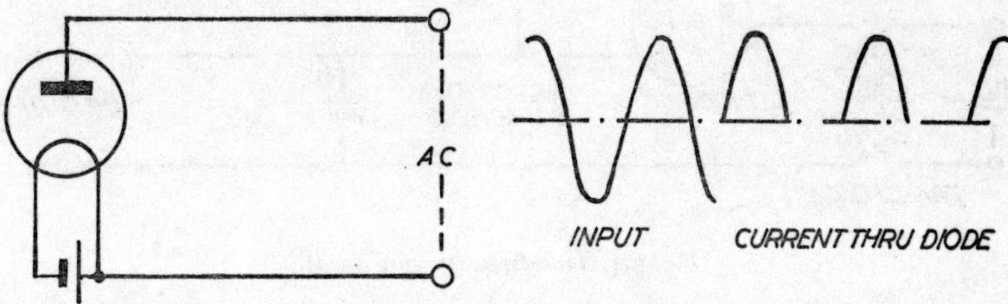

Fig. 5.3. *Diode connected to AC supply (left). Rectification of input (AC) current by diode (right).*

these will attract electrons from the cathode; but shut off and not conduct when negative half cycles are applied to the anode, since these will *repel* electrons. The anode current passed on positive half cycles will be proportional to the applied voltage, as before. The result is that with AC working the diode will conduct pulses of current in one direction only, or change the applied AC into pulsating DC (Fig. 5.3). This is known as rectification, and is widely applicable for changing AC into DC, and for detecting radio signals.

A very simple way of changing AC into DC, and at the same time changing the voltage, is to use a transformer in conjunction with a diode. This principle can be used for making a battery charger or a mains operated DC power supply for a model railway or model car layout. The required step-down in voltage is given by the ratio of the transformer windings, the secondary winding then acting as the feed or high tension supply for the diode. At the same time the transformer can incorporate another smaller secondary winding to provide the filament voltage for the diode and dispense with the need for a separate filament battery. It does not matter that this filament supply will be low voltage AC as we are only concerned with its heating effect.

The basic circuit is then as shown in Fig. 5.4. The output will be pulsating DC with the voltage stepped down according to the transformer winding ratio. If a smoother output is required, then the addition of a

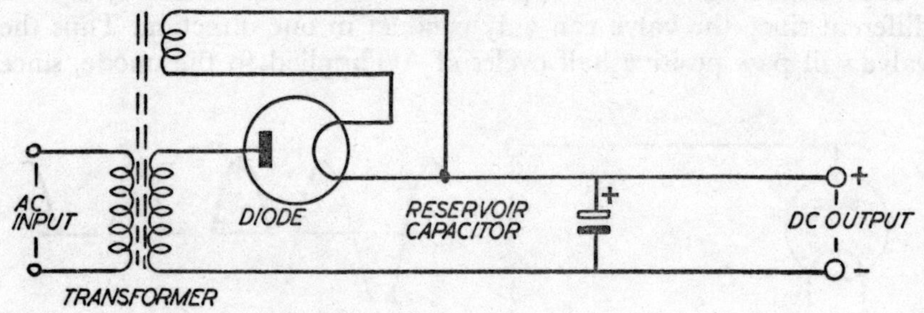

Fig. 5.4. *Transformer/rectifier circuit.*

reservoir condenser across the output will do the trick. This is because the capacitor will store up charge during each pulse of DC and discharge during each subsequent cut-off period, producing a more continuous DC flow.

The output DC will still be "ripply", however, which may not be suitable for some purposes. This can be overcome by using two diodes instead of one, connected so as to work on alternating half cycles. Thus when one diode is cutting off the other is conducting. The result is a continuous DC output (Fig. 5.5).

In practice this is a very convenient solution for valves are made incorporating two identical, but separate valve units, in a single envelope. Thus a double diode is, physically at least, a single valve which can be mounted on a standard valve base, but as far as its internals are concerned, it comprises two separate diode valves for working. The complete circuit for a power pack using a double diode is shown in Fig. 5.6, and this will give a very smooth DC output, almost if not completely free from AC ripple. In this circuit the smoothing capacitor is again shown, followed by a further smoothing filter comprising a 1,000 ohm

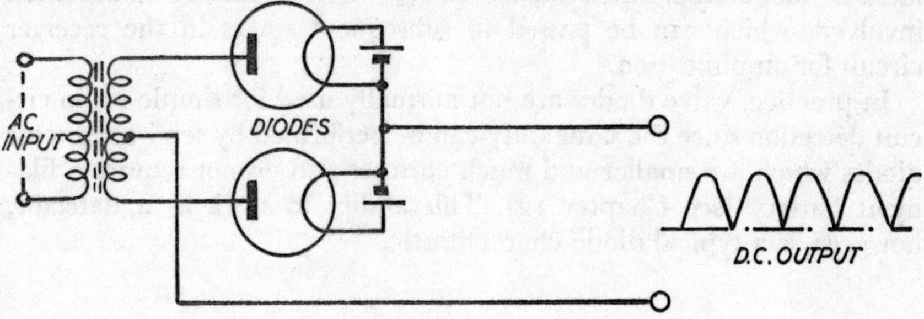

Fig. 5.5. Rectifier circuit giving continuous DC output.

resistor and a second capacitor. These components could be eliminated if freedom from AC ripple is not an essential requirement.

For further information on power packs, and transformer winding, etc., see Chapter 9.

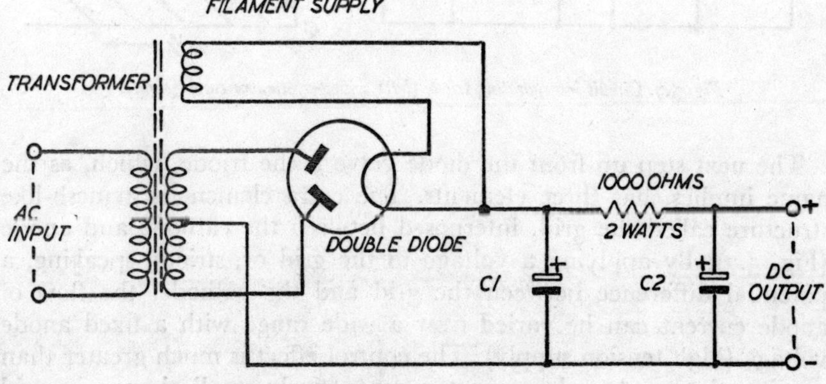

Fig. 5.6. Power pack using a double diode.

As a detector in radio circuits the diode works in a similar fashion. The input in this case is a high frequency AC signal which may be modulated or varied in amplitude by a lower frequency signal imposed on it. The diode chops off all the negative half cycles of the signal to pass a "chopped" high frequency current plus either a steady or varying DC component, depending on whether the original signal is unmodulated or modulated. In effect, therefore, it detects the DC component involved, which can be passed to subsequent stages in the receiver circuit for amplification.

In practice, valve diodes are not normally used for simple radio circuit detection since the same duty can be performed by semi-conductor diodes which are smaller and much cheaper and do not require a filament battery (see Chapter 12). This ability to work as a detector, however, is a typical diode characteristic.

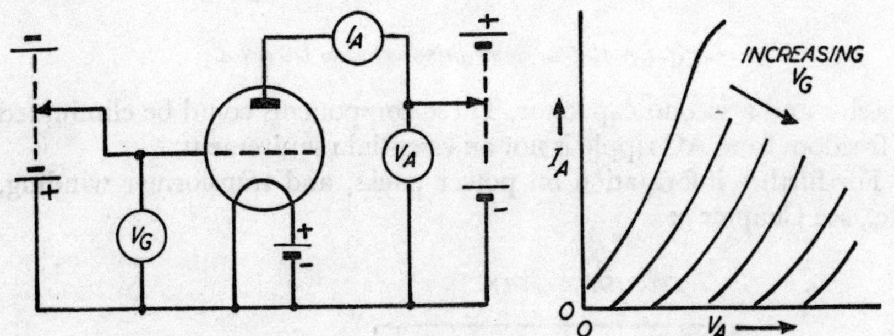

Fig. 5.7. *Circuit incorporating triode (left) with performance curve (right).*

The next step up from the diode valve is the triode, which, as the name implies, has three elements. The extra element is a mesh-like structure called the grid, interposed between the cathode and anode (Fig. 5.7). By applying a voltage to the grid or, strictly speaking, a potential difference between the grid and the cathode, the flow of anode current can be varied over a wide range with a fixed anode voltage (high tension supply). The control effect is much greater than varying the anode voltage as a comparatively small change in grid voltage can produce a large change in anode current.

This gives a triode the characteristics of being able to act as an amplifier, or generate a large voltage by a small one; and the amplification factor of a triode is a measure of how many times the grid voltage is better than the anode voltage in changing the anode current. A figure of about 20 to 30 is quite common.

A basic triode amplifier circuit is shown in Fig. 5.8. The input is a relatively weak AC signal, applied directly to the grid of the valve. The resulting changes in grid voltage produce a corresponding, but much greater change in anode voltage and thus in the anode current flowing through the valve. The input signal has been multiplied or amplified. The actual degree of amplification achieved (represented by the ratio of the output voltage amplitude to the input voltage amplitude), is called the *gain* of the amplifier circuit. This will always be less than the amplification factor of the valve itself, but may approach this figure if the anode load resistance is high.

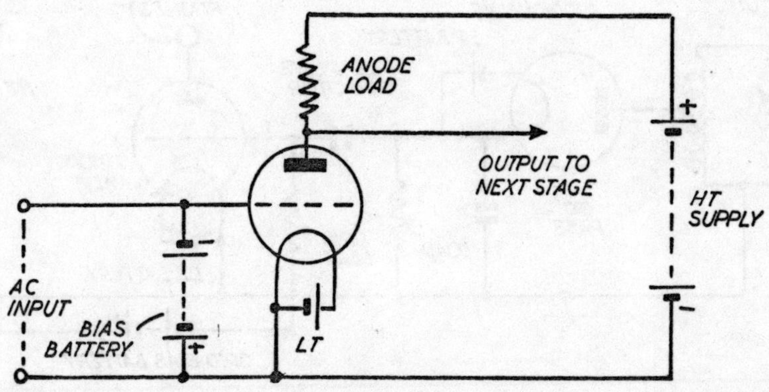

Fig. 5.8. Basic triode amplifier circuit.

Mutual conductance, which is another significant performance factor, is related to the amplification factor and the anode resistance of the valve. The latter quantity, also known as the internal resistance of the valve, is defined as the change in anode voltage divided by the change in anode current which it produces. The relationship between the three factors is then:

$$\text{mutual conductance} = \frac{\text{amplification factor}}{\text{internal resistance (anode resistance)}}$$

It will be noticed that the basic circuit calls for a further battery connected to the grid. The purpose of this is to apply a negative bias voltage to the grid to ensure that whatever the variations in input voltage applied to the grid the resultant grid voltage (which is the combination of the bias voltage and input voltage) always remains within the "straight line" operating part of the characteristic curve of the valve. The bias voltage also has to be large enough (negative) to ensure that the grid voltage can never become positive with any applied signal as this could damage the valve through an excessively high anode current being developed. Thus a correct grid bias voltage is an essential feature of a triode working as an amplifier.

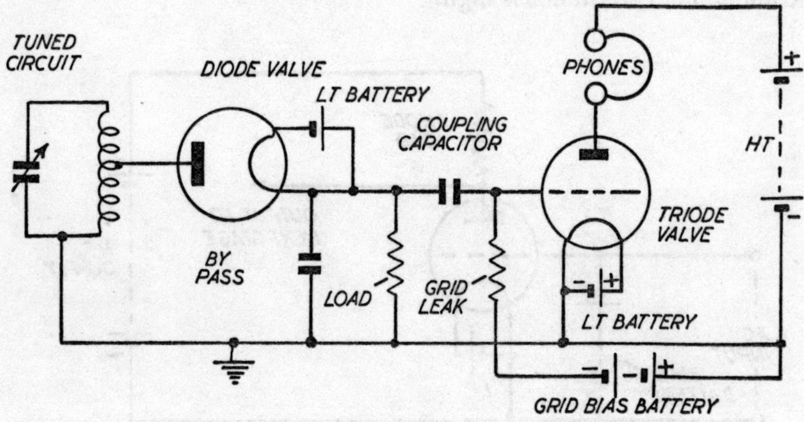

Fig. 5.9. Circuit for demonstrating simple diode (detector) and triode (amplifier) working.

A practical circuit for a triode amplifier used as part of a simple radio receiver circuit following a diode detector is shown in Fig. 5.9. This is an experimental rather than a realistic circuit, for the same results can be achieved more simply with semi-conductors instead of a diode and triode valve, but it does make a circuit for demonstrating simple diode (detector) and triode (amplifier) working. The anode load is provided by the headphones.

In addition to the grid bias battery of specified voltage (according to the type of triode), a resistor is also connected to the grid and is termed the grid leak.

The grid bias battery, incidentally, provides only bias, not current and thus has a very long life. In this way the value of the grid leak does not affect the bias voltage. With certain types of valves designed to operate on a very low anode current (W. 17, for example) it is possible to eliminate the grid bias battery entirely in this type of circuit, the necessary bias being obtained directly from the high tension battery connection through the grid leak. In this case the value of the grid leak resistor will have a marked effect on the volume heard in the head-phones.

Another method of eliminating the grid bias battery is to connect the grid leak resistor directly to the negative side of the low tension battery

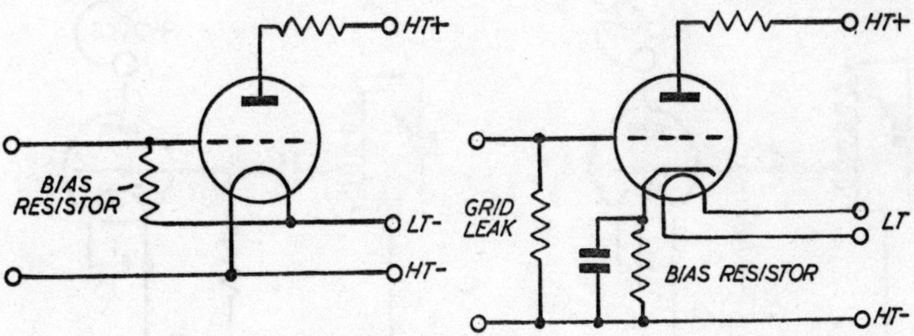

Fig. 5.10. *Two methods of providing bias. In the drawing on the left, the grid leak resistor is connected directly to the negative side of the low tension battery. Bias can also be obtained by using a bias resistor (right).*

(Fig. 5.10). This provides that the grid is more negative than the average filament potential—the difference being the effective grid bias. There are also rather more complicated ways of providing bias from the negative side of the high tension battery when applied to two or more stages. In the case of radio receiver circuits using indirectly heated valves (mains valves), bias can be obtained by using a bias resistor as shown in the second diagram of Fig. 5.10. The value of bias resistor required is calculated from the simple formula:

$$\text{bias resistor} = \frac{\text{grid bias volts required}}{\text{DC anode current}}$$

The triode valve can also be used as a detector when, in a suitable circuit, it can act as an amplifier at the same time (Fig. 5.11). Used as a basic detector (anode bend detector) a fixed bias is employed of just the right value to cut off the anode current when no input signal is present. Thus when worked by an input signal only positive half cycles are passed by the valve conducting. In this respect it is working the same way as a diode, but there is also amplification of signal due to the triode characteristics.

In the second circuit shown in Fig. 5.11 no standing bias is provided so that in the absence of a signal both the grid and cathode are at the

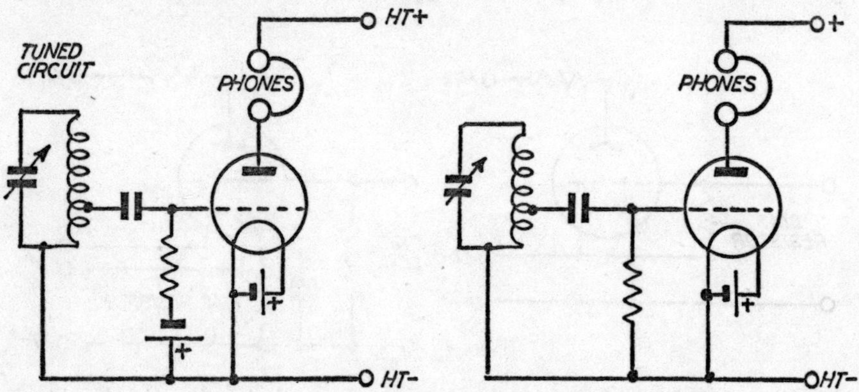

Fig. 5.11. *Two simple one-valve receivers. Left, the anode bend detector; right, a circuit where the triode valve acts as both a detector and amplifier.*

same potential. The grid and cathode then behave as a diode in the detector circuit, with the grid acting as the diode anode. When a signal flows in the detector circuit the DC component introduced in the resistor provides a negative self-bias to the grid to promote working of the valve as a triode amplifier. Both the AF and RF components of the signal are, in fact, amplified, but the introduction of a capacitor between anode and HT negative acts as a short-circuiting path to remove unwanted RF.

Both the circuits shown, incidentally, are complete radio receivers, when coupled to a suitable tuned circuit (see Chapter 12). Performance, however, is likely to be poor except in areas of good reception. It can be improved by following with a further stage of amplification; replacing the phones with an anode resistor of suitable value for the valve used and coupling the bottom of the anode to the next amplifier stage with a capacitor of 8 to 10 mf to any of the triode amplifier circuits previously described.

Triode detectors, however, are not all that good for working in radio circuits since they tend to introduce distortion, and the greater the input signal the greater the distortion is likely to be. Diode detectors are very much better in this respect and usually preferred. Triodes find their main application in amplifiers (see also Chapter 10) and oscillators (see Chapter 11).

Further developments of the triode are known generally as screen

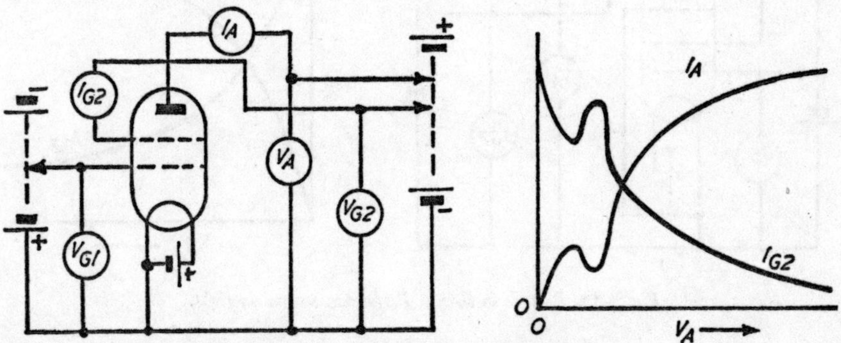

Fig. 5.12. Tetrode in circuit. Performance curve on right.

grid valves, of which the main types are the tetrode (Fig. 5.12) and pentode. The tetrode has a second (screen) grid inserted between the conventional control grid and the anode which acts as an electrostatic shield and reduces the grid-anode capacity to a negligible value. This can be of specific benefit in amplifiers where it is necessary to prevent oscillation. In a practical circuit the screen grid is connected to the cathode through a low impedance path, such as a by-pass condenser.

The pentode (or five element valve) goes one stage further in introducing yet another screen called the suppressor grid, between the screen grid and anode (Fig. 5.13). This grid is connected directly to the cathode and further improves the operating characteristics of the valve. Basically, in fact, whilst the mutual conductance of a tetrode or pentode are of the same order as that of a triode, both the amplification factor and anode resistance of these screen grid valves is very much higher. They are thus capable of a superior performance as amplifiers, with a gain of 100 to 200 readily attainable.

A little thought will show that although tetrodes and pentodes seem a much more complicated type of valve, they still work, basically, in the same manner as a triode, but offer some improvement in performance.

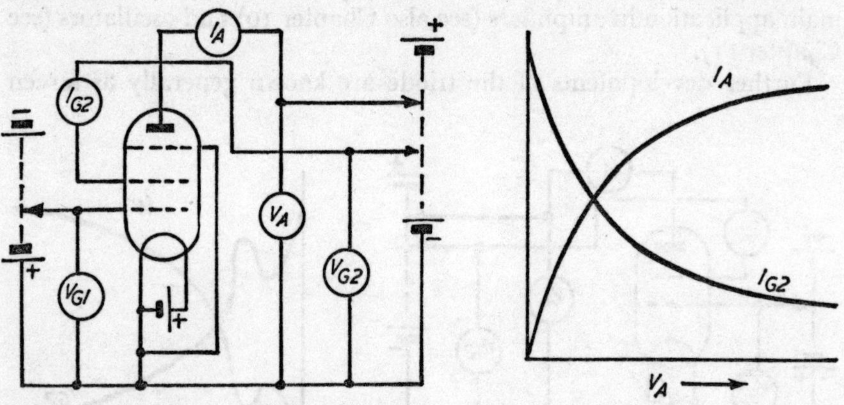

Fig. 5.13. Pentode in circuit. Performance curve on right.

The only complication as far as circuit construction is concerned is that there is one (or two) more leads from the valve to connect up—that is all! For connection purposes, valves are invariably mounted in valve bases of matching type and connections made to the appropriate tags on the valve base. These are related by number to corresponding connections with the innards of the valve, and usually presented in the form of a simple diagram.

CHAPTER 6

SEMICONDUCTORS (DIODES AND TRANSISTORS)

S EMICONDUCTORS represent an entirely different type of circuit element to thermionic valves, but with many similar working characteristics. Physically they are quite different; they are solid state devices; can be made much smaller and, lacking the necessity of an evacuated envelope to contain their elements, are very much more robust.

They are, in fact, as the name suggests made from materials such as germanium and silicon which have properties between that of an insulator and a conductor. The ability to act as either an insulator or a conductor can be modified by the introduction of minute traces of impurity in the material, and also tends to change with temperature. This latter characteristic places a limit to the working temperature of the device, which in turn means the amount of current it can carry without overheating and suffering a drastic change in characteristics.

The theory as to why these particular materials should have semi-conductor properties is relatively complex, but for the purpose of understanding the application of such devices it is only necessary to appreciate that there are two basic types of semiconductors with char-acteristics governed by their atomic structure. One, known as the P-type, has an atomic structure with a wandering positive charge which has the effect of yielding a material with a number of electronic "holes". The second form, known as the N-type, has a surplus electron in its atomic structure, although the material is electrically neutral as a whole. N-type material will act like any normal conductor when connected to a battery.

To yield practical properties from these materials the two must be combined in a single component—for example a piece of P-type joined to a piece of N-type. We then have a component element shown dia-grammatically in Fig. 6.1, with "holes" on one side of the junction and random electrons or negative charges on the other side. On either side of the junction there will be a depleted region in the N material where some "holes" have migrated into the N region and absorbed a corresponding number of wandering electrons. There is a depleted region in the P material where electrons have wandered across the junction to fill a corresponding number of "holes". A state of equilibrium will quickly be reached where "holes" moving into the N material have left the P material with an overall negative charge tending to hold the remaining "holes" more firmly in place; and similarly an overall positive charge developed in the N material to hold further electrons back from the junction. Nor will connecting the two ends to complete an external electrical circuit have any effect. This will only be equivalent to connecting the N material to the P material to try to pull electrons out of the P material where only "holes" exist.

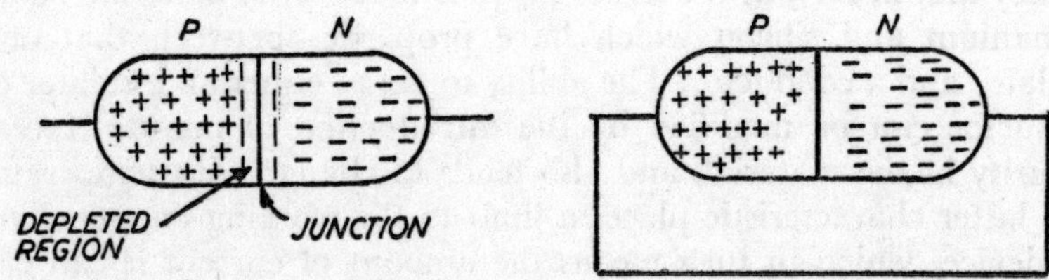

Fig. 6.1. *Diagrammatic representation of the electronic structure of a diode.*

Completing the external circuit with a battery, however, will certainly have an effect (Fig. 6.2). If the positive of the battery is connected to the P material (and the negative to N) the battery is feeding the N material with more electrons and at the same time pulling electrons out of the P material (equivalent to pulling more holes in the P material). The result is that the device acts as a normal conductor with a free flow of electrons across the junction. The only resistance to flow is the intrinsic resistance of the P and N material, which is very low. Thus current flow will be initiated on about a quarter of a volt in the case of germanium material and roughly twice this figure in the case of silicon material.

If the battery is connected the reverse way round—battery negative to P—the device now offers extremely high resistance to current flow. This is because the electrons fed to the P material are giving it a stronger hold over its "holes"; and the positive connection to the N material is increasing its hold on its electrons. The result is to increase the depletion area substantially with a high barrier effect opposing current flow.

The P–N element, therefore, acts as a conductor connected one way round, and an insulator the other way round, that is as a rectifier. It is, in fact, a diode with similar characteristics to a valve diode, but on a different voltage level. Its operating characteristics are not identical for the semiconductor diode is subject to a certain leakage current in the non-conducting direction, with its operating characteristics as shown in Fig. 6.3. It can, however, be used for similar purposes as a circuit element—for rectification and detection, for example.

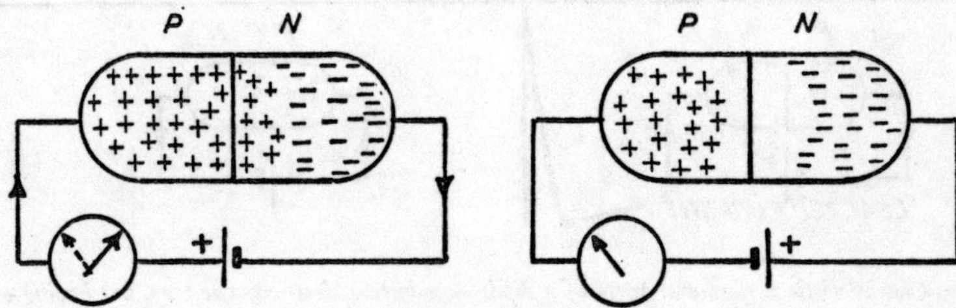

Fig. 6.2. Electron flow with diode connected to a battery—current flows with forward bias (left); but no current flows with reverse bias (right).

The leakage current will be an inherent feature of the diode material and construction. It can also occur over the surface of the element and vary with the cleanliness of the surface. That is why semiconductor diodes (and transistors) are hermetically sealed to keep them clean and stabilize leakage current characteristics.

Semiconductor diodes of this type are made in a wide variety of forms and sizes. One of the most common types is the tiny wire-ended, glass-encased point-contact type where the junction is formed by fusing a wire to the surface of either a P or N type crystal to form a P–N junction. This provides a high junction resistance compared with a true junction diode made from a P and N crystal joined together, but is a simpler and cheaper form of construction. Point-contact diodes are suitable for most radio work, etc., although junction diodes are invariably employed where higher reverse voltage ratings or greater current carrying capacity is required.

Examples of the use of the simple point-contact diode for radio circuits will be found in Chapter 12. Its use as a rectifier will be seen in a number of various designs throughout this book.

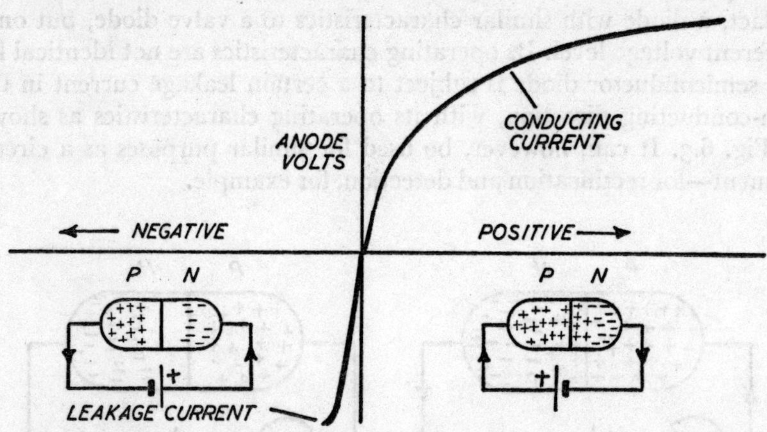

Fig. 6.3. *Characteristic performance curve of a diode—conducting electricity one way, but having only a very small leakage current connected the other way round.*

The transistor is a rather more sophisticated semiconductor device involving a three element combination—either P–N–P or N–P–N—Fig. 6.4. Both work in exactly the same way, except that the direction of current flow is reversed.

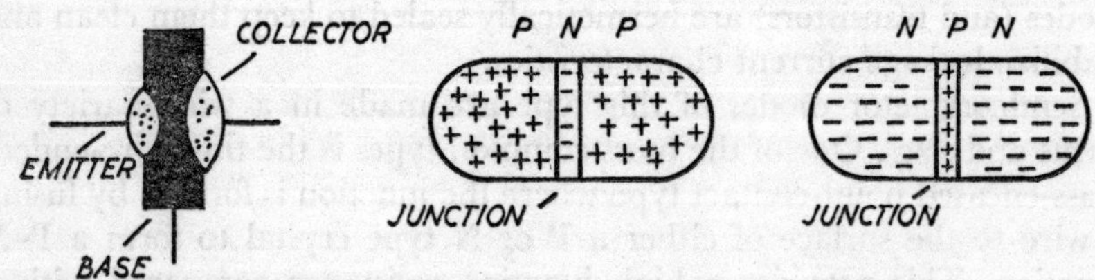

Fig. 6.4. *Diagrammatic representation of the electronic structure of a transistor.*

The three parts involved, with their separate connections, are known as the collector, emitter and base; the whole forming together what is, in effect, two diodes with the base as a common element. A separate external battery can then be applied to each half, as shown in Fig. 6.5, to provide "bias", and work each half as a diode circuit.

Considering first the collector and base as one diode circuit, connecting with reverse bias will produce only a small leakage current flow through the base-collector junction. Completing the emitter-base circuit also with reverse bias will have a similar effect in that "diode" circuit. If, however, the voltage applied to the emitter-base circuit is reversed (i.e. forward bias), "holes" are injected into the N region

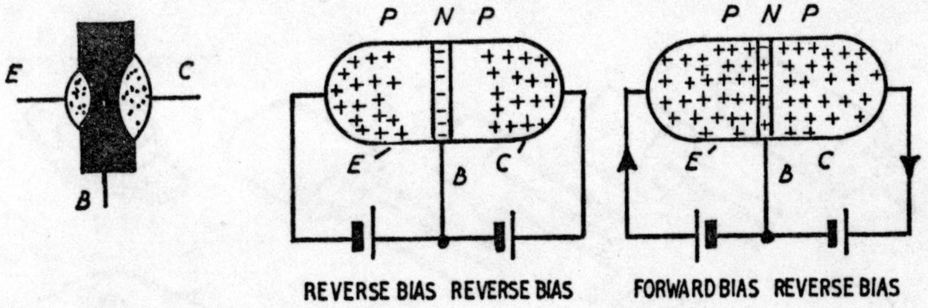

Fig. 6.5. *Characteristic performance of a PNP transistor when connected to two batteries to complete the two external circuits.*

(base) where they will tend to diffuse through into the right hand P region. The overall effect is a substantial increase in current flow through the base-collector circuit. The three elements are also defined by the P region from which "holes" are injected into the N region as the emitter; the N region as the base; and the P region which collects the flow as the collector.

In working transistor circuits, forward bias is applied to the emitter-base circuit to provide a low resistance path. In other words the positive battery terminal will always connect to the emitter on a PNP transistor; and the negative terminal of the battery will always connect to the emitter on an NPN transistor. Reverse bias is applied to the base-collector circuit so that for a PNP transistor the negative terminal of the battery connects to the collector; and the positive terminal of the battery connects to the collector in the case of an NPN transistor. Collector and base will require the same polarity, which will always be opposite to the emitter polarity.

Physically the size and shape of transistors can vary considerably, although all common types have three separate leads or wires emerging from the bottom of the transistor (Fig. 6.6). These three wires are identified as the collector nearest to a coloured dot or mark on the bottom or side, the other end wire being the emitter and the wire in between the base. The base and emitter wires are also normally closer together than the base and collector. On some older types of transistor the three wires emerge in triangular configuration, with the collector again marked by a coloured dot, as shown in Fig. 6.6.

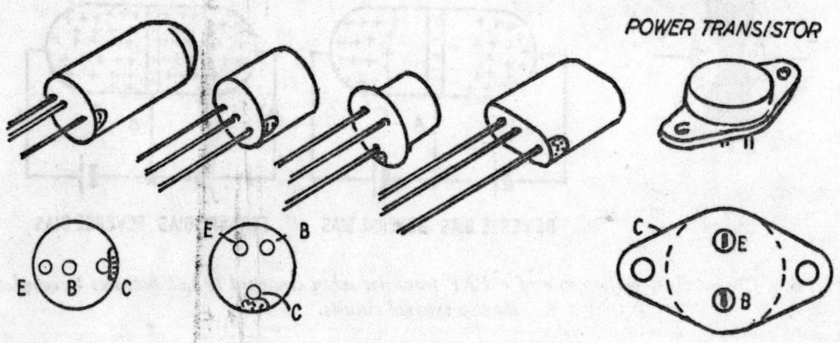

Fig. 6.6. *Types and shapes of transistors.*

In the case of "power" transistors—those constructed to handle fairly heavy currents—the shape is usually flat, with an elongated mounting base. Only two leads protrude from the bottom of the transistor, marked "E" for emitter and "B" for base. The collector is internally connected to the mounting plate of the transistor and so the collector connection is made to this. In order to dissipate heat, which could otherwise interfere with the working of the transistor, power transistors of this type are normally bolted in place to a "heat sink". In simple circuits an adequate heat sink can be provided by a sheet of 16 s.w.g. aluminium of adequate size.

Correct identification of leads is essential in wiring up transistors as wiring up with the wrong polarity may destroy the component. For experimental circuit construction transistor holders can be used, into which the three transistor leads plug. This is particularly useful where it may be necessary or desirable to change the transistors used in the circuit. For permanent circuits, assembly is usually on a printed circuit panel with the transistor soldered in place by its leads.

On circuit drawings a transistor is identified by the symbols shown in Fig. 6.7. The base is normally represented by a thick, short line, with lines angled from it on one side representing the collector (plain line) and emitter (line with arrowhead). The direction of the arrowhead also identifies the type of transistor (PNP if the arrowhead points towards the base; NPN if the arrowhead points away from the base). The transistor symbol may be drawn with or without an enclosing circle. There is also a variation used where collector and emitter are drawn on opposite sides of the base, in line.

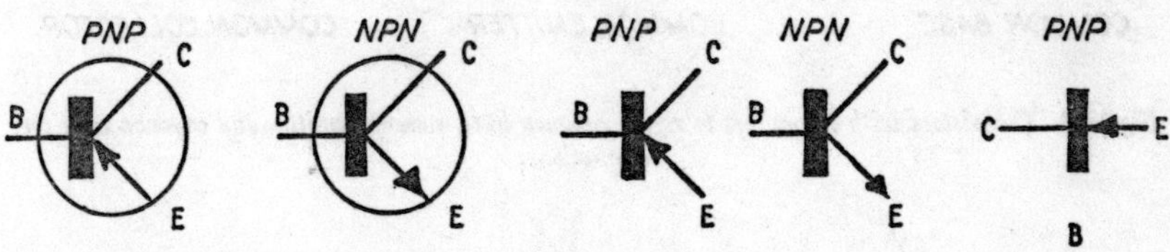

Fig. 6.7. Symbols for transistors.

Practical circuits for investigating the characteristics of transistors are useful both to gain familiarity with what transistors can do, and also as a means of proving the characteristics of individual transistors. The latter can vary considerably, even for transistors of the same type number of specification. In some circuits close matching of transistor characteristics is necessary, and this can only be done by testing in a working circuit.

Since a transistor embodies two diodes with only three leads it follows that one lead, or one element of the transistor, must be common to both circuits. There are three possibilities—common base, common emitter and common collector (Fig. 6.8). The characteristics on both sides of the circuits can be determined independently; and also the combined or transfer characteristics of the transistor.

No special instruments are required for transistor testing and measurement other than conventional ammeters and voltmeters, except that in certain cases the order of current or voltage to be measured is quite small. Thus accurate measurement of leakage currents requires the use of a microammeter; and measurement of diode characteristics may require the use of a voltmeter reading to only 1 volt full scale. In this case, for accurate measurement a valve voltmeter or very high resistance voltmeter needs to be used (see Chapter 3).

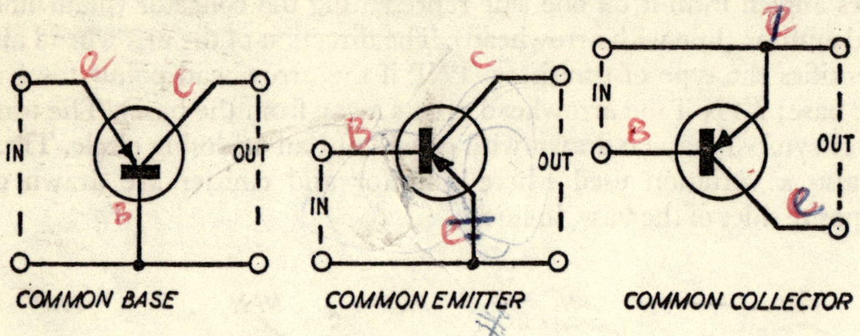

Fig. 6.8. *Transistors with (from left to right) common base, common emitter and common collector connections.*

The same circuit can be used to test any type of transistor, the important thing being that the maximum voltages and currents which can be developed cannot exceed the limiting values for that transistor being used in the circuit. These are always specified by the transistor manufacturer. The component values employed in the following circuits are based on the OC 71 transistor, where the following limiting values apply:

	Symbol	*Limiting value*
COLLECTOR VOLTAGE	V_c	20 volts
COLLECTOR CURRENT	I_c	10 milliamps
EMITTER CURRENT	I_e	12 milliamps
BASE CURRENT	I_b	2 milliamps
BASE-EMITTER VOLTAGE	V_{be}	10 volts

The basic circuit for the transistor in common base configuration is shown in Fig. 6.9 with these symbols included. Fig. 6.10 shows the same circuit with actual component values required.

With SW1 closed and SW2 open, this circuit measures the forward emitter-diode characteristics of the transmitter. This is not particularly instructive, except that by varying R_1 the emitter voltage can be changed and the corresponding emitter current noted. The relationship between V_e and I_e will be as shown in Fig. 6.11.

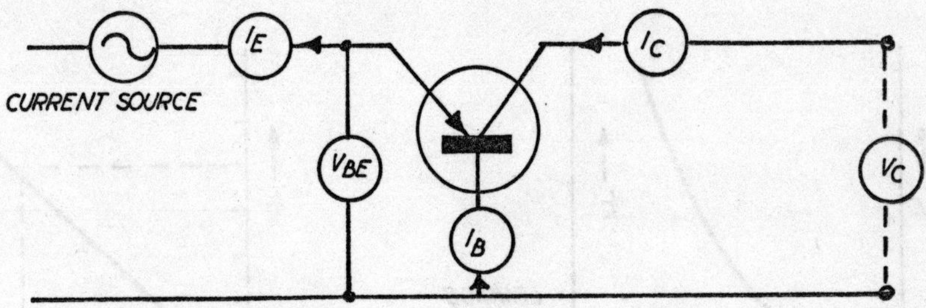

Fig. 6.9. *Basic circuit for transistor in common base configuration.*

With SW1 open and SW2 closed, the circuit is measuring the reverse collector-diode characteristics, and the corresponding collector current is known as the leakage current. This will be quite low with a good transistor—for example, of the order of 4 microamps—and so will only be detected by a sensitive ammeter (microammeter). It will probably

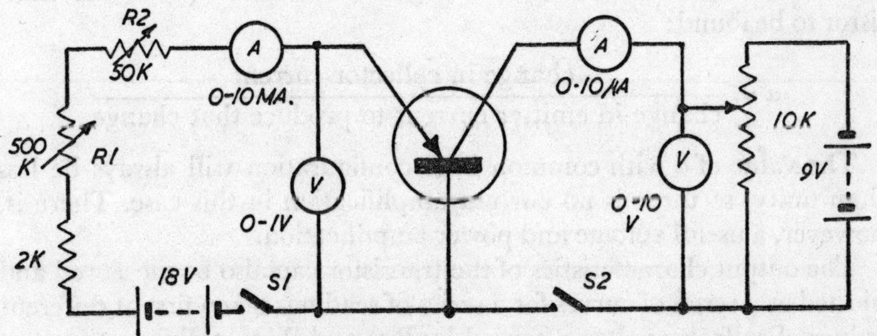

Fig. 6.10. *Circuit as in Fig. 6.9 with component values required.*

be impossible to detect any change in leakage current with a change in collector voltage, which is produced by varying R_2, although in fact there is such a change (see Fig. 6.11).

With both SW1 and SW2 closed, the circuit measures the base transfer characteristics of the transistor. The emitter current can be varied by means of R1, when it will be found that the emitter current will vary in almost exact proportion (see Fig. 6.11). For accurate measurement the collector voltage must be kept the same throughout the experiment

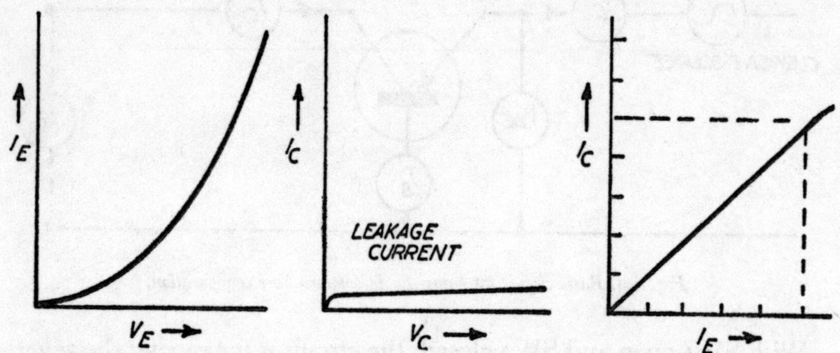

Fig. 6.11. Characteristic performance curves of a transistor in common base configuration.

—for example at 4·5 volts by adjusting R2 as necessary. A plot of I_c against I_e then enables the current gain or "alpha" (α) of the transistor to be found:

$$\alpha = \frac{\text{change in collector current}}{\text{change in emitter current to produce that change}}$$

The value of α with common base configuration will always be less than unity, so there is no current amplification in this case. There is, however, a useful voltage and power amplification.

The output characteristics of the transistor can also be measured and plotted as a series of curves for a series of readings taken first at different values of collector voltage (varied by R3) and then at different emitter currents (varied by R1). The resulting family of curves will be like those shown in Fig. 6.11 and are similar to those produced by a pentode valve.

The same series of experiments can be repeated with the transistor connected in common emitter configuration (Fig. 6.12), although only the reverse-diode, transfer and output characteristics have any significance. The reverse collector-diode characteristics are determined with

SW1 open and SW2 closed when the collector voltage can be varied by R3 and the corresponding collector current measured. This again is a leakage current and will increase sharply at first and then taper off to about 100 microamps (see Fig. 6.13).

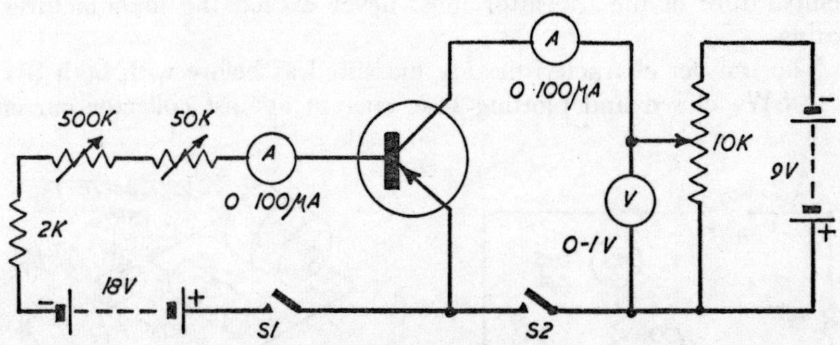

Fig. 6.12. *Transistor connected in common emitter configuration.*

It will be noticed that this is substantially larger than with the same transistor in common base configuration. Also with this configuration the leakage current is very temperature dependent. Grasping the transistor between the thumb and forefinger will cause it to absorb heat, when it will be seen that the leakage current will rise. In a practical circuit if the leakage current increases for any reason the total current

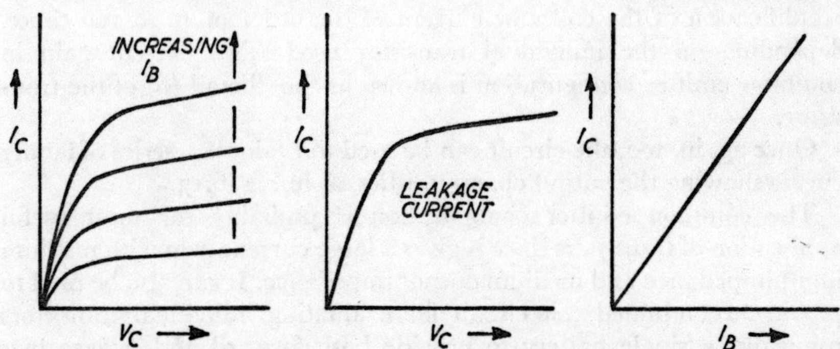

Fig. 6.13. *Characteristic performance curves of a transistor in common emitter configuration.*

through the transistor will increase causing a further increase in leakage current. There can come a point where this process becomes accelerated and the transistor "runs away" and will destroy itself. With common emitter configuration, therefore, it is important that the transistor never be allowed to approach "run away" conditions. For example the temperature of the transistor must never exceed the manufacturer's rating.

The transfer characteristics are measured as before with both SW1 SW2 closed and plotting base current against collector current

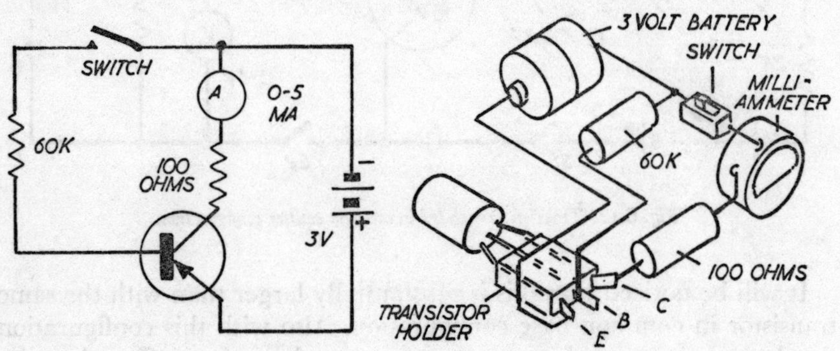

Fig. 6.14. Simplified test circuit for evaluating individual transistors. Theoretical circuit on left and physical circuit on right.

(Fig. 6.13). It will be seen that this time there is a substantial gain or amplification of the collector current of the order of 20 to 100 times, depending on the individual transistor used. This current gain in common emitter configuration is known as the "beta" (β) of the transistor.

Once again, too, the circuit can be used to evaluate a series of family curves showing the output characteristics, as in Fig. 6.13.

The common emitter configuration is probably the most useful application of transistors since it gives a large current gain with medium input impedance and medium output impedance. It can also be used to provide a simplified test circuit for evaluating individual transistors using just a single battery to provide both forward and reverse bias (Fig. 6.14). This is possible since the collector voltage is greater than the base voltage with respect to the emitter, hence a difference in potential will exist between base and collector representing a reverse bias.

The values of the resistors specified for this circuit are such that the numerical value of gain is 20 times the reading obtained on the milliammeter. Thus simply plugging in a transistor and closing the switch enables the "beta" of that transistor to be obtained at once as 20 times the meter reading.

This simple circuit can also be extended by replacing the meter with a two-point socket. A two-point plug is then wired to the milliammeter to plug into the circuit for "beta" measurement. If a 1–100 micro-ammeter is plugged into the meter point instead of the milliammeter and the switch left open, this meter will read the leakage current of the transistor (in common emitter configuration). Further, if phones are plugged into the socket instead of a meter and the switch closed, the inherent noise developed by the transistor can be listened to. Thus the circuit provides a ready means of evaluating individual transistors on the basis of "beta", leakage current and noise.

The common collector configuration can be evaluated in a similar manner to the methods already described above, but since it only has a very limited practical application it will not be described in detail. Suffice it to say that it produces a large current gain of the same order as the common-emitter circuit, a high input impedance and a low output impedance. Leakage current is of the same order as that with common emitter configuration. The common collector configuration is basically the equivalent of a "cathode follower" in a thermionic valve circuit.

CHAPTER 7

PHOTOCELLS AND SOLAR CELLS

A LIMITED number of materials are photosensitive or sensitive to light. Photographic emulsions form one group of practical photosensitive materials; another group combine photosensitivity with certain electrical phenomena and are generally described as photoelectric devices, or more conveniently photocells.

Photocells may act rather like radio valves in that a change in illumination, or of light falling on them, produces a change in their properties of conduction as a component in an electric circuit; or where light falling on the cell can actually produce or generate electricity in a

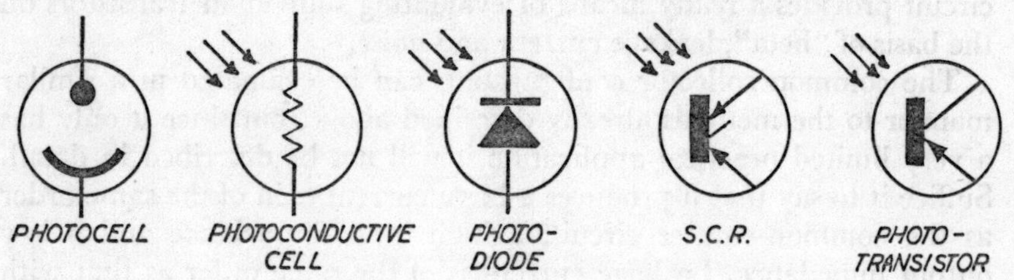

Fig. 7.1. *Symbols for photo-electric devices.*

circuit where none existed before. The former are correctly described as *photoconductive* cells, and the latter as *photovoltaic* cells. There is also a third type known as a *photo-emissive* cell which are rather like conventional diodes both in construction and performance and for that reason are commonly called *photodiodes*. They have the property of one-way-only transmission of current (or a rectifying action), like an ordinary diode, but this function is activated by light falling on the cell. Then there is also the *phototransistor* or light-sensitive transistor which combines a similar photoconductive function combined with the amplifying ability of a transistor.

All these devices have their separate symbols (see Fig. 7.1). This also includes a relatively new device known as an SCR or light-sensitive silicon controlled rectifier. Basically this looks rather like a transistor but generally has a transparent top. It combines the features of a high current switch action with the light sensitivity of a photoconductor.

The working of a photovoltaic cell is the simplest to understand. Basically such a cell normally consists of a thin film of a semi-conductor material such as selenium or silicon deposited on a steel or copper plate. The surface of the semiconductor is then covered with a film of noble metal, such as gold, which is so thin that it is transparent to light. A

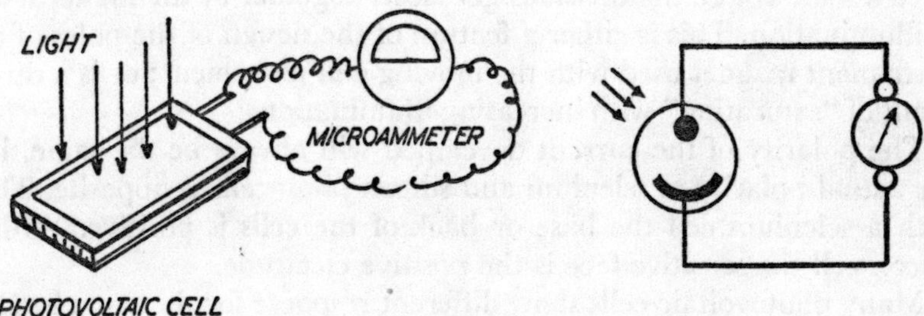

Fig. 7.2. Photovoltaic cell connected with microammeter forming simple light meter. Theoretical circuit on right.

metal ring over this transparent coating then forms one contact of the cell, and the steel or copper plate the other.

In practice, such cells normally look like metallic wafers and may be produced in a variety of shapes and sizes from about $\frac{3}{32}$ sq. in. surface area up to 12 sq. in. surface area (selenium cells); or from about $\frac{3}{32}$ sq. in. up to about $1\frac{1}{4}''$ diameter (silicon cells). Thickness is of the order of ·002″ to ·006″.

Light falling on to the photosensitive surface of such cells has the effect of liberating electrons in the boundary layer of the sandwich construction with the result that if connected to an external circuit, an electric current will flow in that circuit. Within limits, the amount of electricity generated is proportional to the amount of light falling on the cell. The current generated is quite small—only a few microamps—but this is sufficient to give a reading on a microammeter (Fig. 7.2) or operate a sensitive moving coil movement. This is the basis of the lightmeter. All you need is a photovoltaic cell and a microammeter, connected as shown in Fig. 7.2, and you have a lightmeter capable of measuring the quantity of illumination falling on the face of the cell. It can only be calibrated, however, with reference to the specific performance curve for that cell; or, more simply, by comparing readings against those obtained from an existing lightmeter (such as an exposure meter).

Strictly speaking, provided the cell is not over-illuminated so that the cell is "saturated", the microammeter reading should be almost directly proportional to the level of illumination. Most lightmeters, however, have a scale where the divisions get closer together at the higher levels of illumination. This is either a feature of the design of the poles of the permanent magnet used with the moving coil movement; or is a direct result of "saturation" with increasing illumination.

The polarity of the current developed will always be the same, but the actual polarity of selenium and silicon photocells is opposite. Thus with a selenium cell the base or back of the cells is positive. With a silicon cell the sensitive face is the positive electrode.

Many photovoltaic cells show different response for the same level of illumination produced by different light sources. Thus whilst a selenium-steel cell will give similar readings for daylight and tungsten bulb light at the same levels of illumination it will under-indicate the light from discharge lamps. A copper oxide-copper cell, on the other hand, will over-indicate daylight with respect to tungsten light, and under-indicate with discharge lamps, but not fluorescent lamps.

There is also a distinction to be drawn between photovoltaic cells which have a low internal resistance (a few thousand ohms only) and a high internal resistance (of the order of megohms). Cells with low internal resistance can be classified as *current* generators; and cells with high internal resistance as *voltage* generators. Since it is also a feature of such cells that the response characteristics are affected by the load resistance, this can govern the choice of type for particular applications. Thus whilst a low resistance photovoltaic cell can generate useful current, both the value of this output current and the linearity of response will fall off with increasing external resistance. Thus any meter movement employed in working the cell in a practical circuit, for example, must have a very low resistance. A high resistance cell, on the other hand, whilst generating a much smaller current is less affected by load resistance and can develop a useful voltage across a high load resistance, which can then be amplified, as necessary.

The silicon cell has a much greater generation capability, particularly in sunlight. Because it works best in sunlight it is often called a solar cell, and solar batteries, comprising a number of such cells connected in series can generate quite worthwhile voltages and currents in strong sunlight. Like any other battery, the voltage output of any photo-voltaic cell can be increased by connecting a number of cells in series; and its current output by connecting a number of cells in parallel.

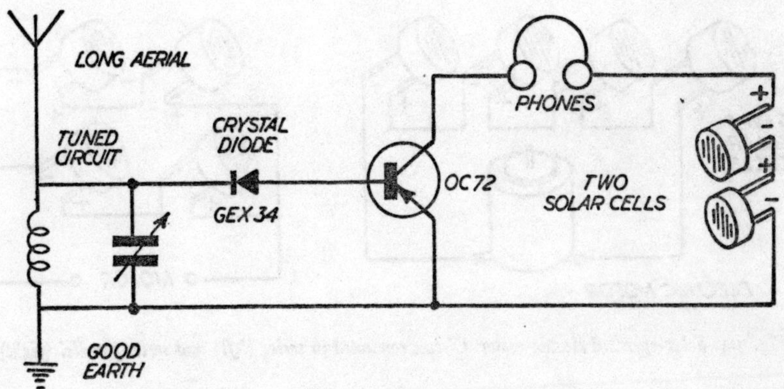

Fig. 7.3. *Solar-operated single-transistor radio receiver.*

A typical silicon solar cell such as currently available is capable of an output of up to 2 volts with a current rating of 10 to 15 milliamps when exposed to strong sunlight. Three or four such cells connected in series can, therefore, supply sufficient voltage to operate a small all-transistor radio receiver (see Fig. 7.3); or even drive a small 3-volt electric motor running light under no load (see Fig. 7.4). In the latter case, however, a series-parallel connection of four or six solar cells would probably make a more effective battery.

Photoconductive cells are quite different in their working properties for they need to be supplied with an external source of electricity and the effect of varying the level of illumination falling on the cell is to give them a variable electrical resistance, the effective resistance changing in proportion to the light level.

Cells of this type often look rather like radio valves, being enclosed in an evacuated glass envelope, although some are flat and disc shaped with a transparent top. All have a specific direction for alignment— a side or face of the cell which must point towards the light source for maximum response.

A basic circuit with a typical performance curve is shown in Fig. 7.5. The electrical resistance of the photocell falls, and thus the current flowing through the circuit increases, with increasing illumination,

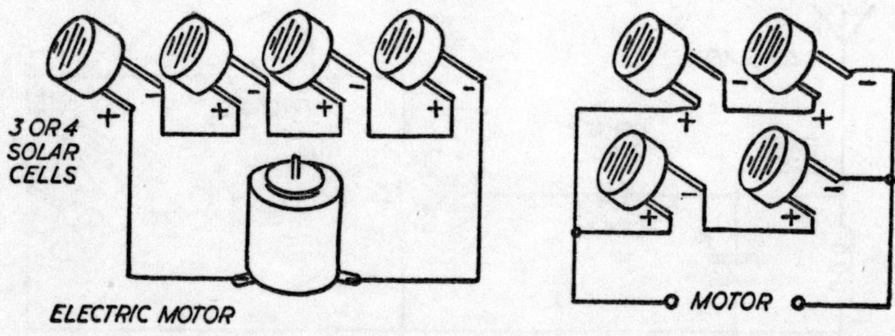

Fig. 7.4. *Solar-operated electric motor. Circuit connected in series (left) and series-parallel (right).*

up to a point where the cell is saturated with light and achieves its minimum resistance value. The current change is not directly proportional to the amount of light, so such a cell is not suitable for measuring different intensities of light. It does, however, distinguish very sharply between a low and a high light level. It can thus act as a very effective switch in providing a substantial change in current in a circuit when the illumination level changes from light to dark, or vice versa. This property is readily adapted to simple alarm and detector circuits, etc. (see Chapter 8). The two main types of cells used for this purpose are the cadmium sulphide cell, which is responsive to visible light, and the cæsium sulphide cell which is highly sensitive to light in the infra-red region.

Photo-emissive cells or photodiodes are rather like diode valves with the cathode formed from an alkali metal with a special coating and a separate anode, sealed in an evacuated glass envelope, or a partially evacuated envelope containing an inert gas. The latter are low voltage types since an excess anode voltage could cause a glow

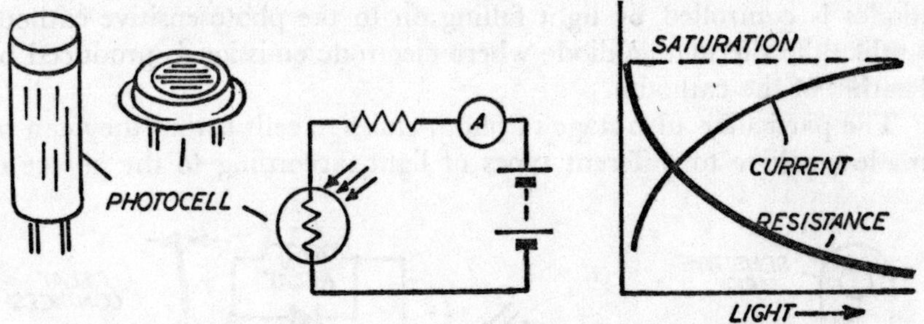

Fig. 7.5. *Basic photoconductive cell circuit with typical performance curve.*

discharge to form and destroy the cell. They are more sensitive than vacuum cells, but less stable.

The main limitation of photodiodes is that the current change realized is usually quite small (compared with photoconductive cells) and so normally they have to be used with one or more amplifier

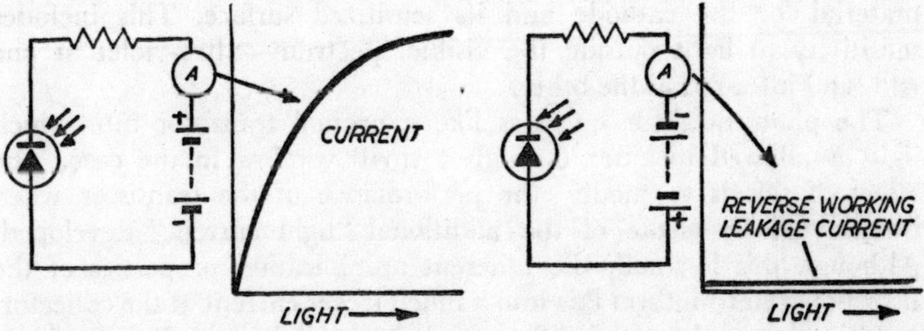

Fig. 7.6. *Characteristic curves of a photodiode showing typical diode action or 'one way' working as a conductor.*

stages in a practical circuit (Fig. 7.6). Sensitivity is improved with the secondary emission cell which employs two cathodes—a primary cathode which receives the light, and a secondary or target cathode which is held at a positive potential to the primary cathode. Electrons from the primary cathode striking the target cathode cause a secondary emission, thus greatly increasing the number of electrons produced and therefore the sensitivity of the cell. Emission of electrons in all photo-diodes is controlled by light falling on to the photosensitive cathode (unlike the thermionic diode where electrode emission is produced by heating of the cathode).

The particular advantage of photo-emissive cells is that they can be made sensitive to different types of light, according to the choice of

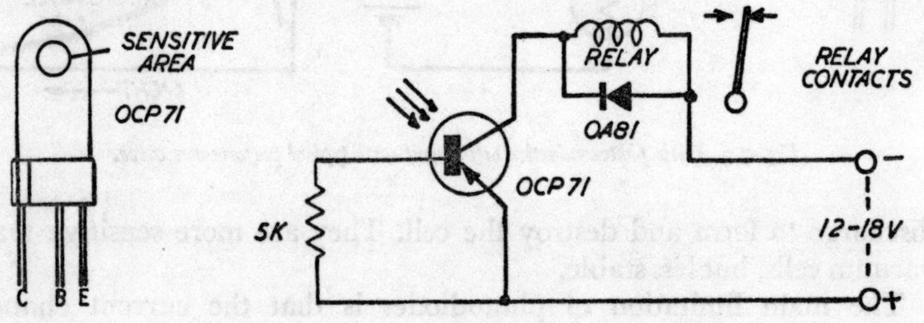

Fig. 7.7. Basic circuit for phototransistor light switch.

material for the cathode and its sensitized surface. This includes sensitivity to light outside the visible spectrum—ultra-violet at one end, and infra-red at the other.

The phototransistor operates like a normal transistor into which light is allowed to enter through a small window in the case. The effect of this is to modify the performance of the transistor when illuminated by virtue of the additional "light current" developed. Although this is small, the inherent amplification properties of the transistor can transform this into a much larger current at the collector, resulting in a substantial difference between light and dark working. Peak response, incidentally, is towards the infra-red end of the spectrum, although phototransistors are sensitive to the whole of the visible spectrum.

A basic circuit for a phototransistor light switch is shown in Fig. 7.7, using an OCP71 transistor. The only other components involved are a sensitive relay and a diode, and a resistor connected between the base and emitter of the transistor to provide stability. A suitable value for this resistor is 5 kilohms. The relay coil resistance should be of the order of 2,000 ohms.

This circuit works as a light-sensitive on/off switch, with the relay contacts themselves forming the switch for an external circuit. Thus when light falls on the transistor the current passed through the relay

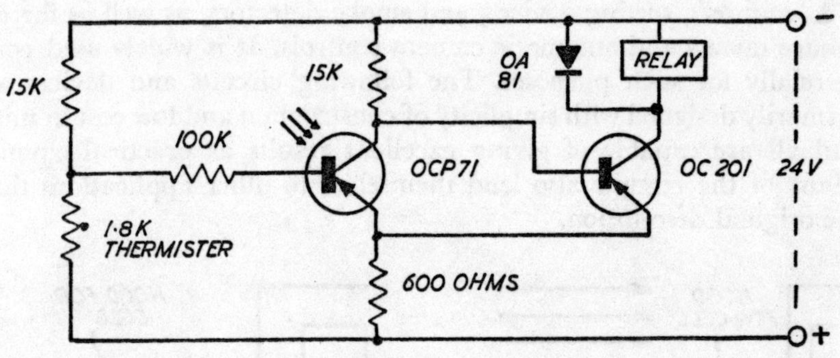

Fig. 7.8. *Improved sensitivity phototransistor light switch.*

is a maximum, pulling the relay in. When the level of illumination falls there will come a point where the corresponding fall in current causes the relay to drop out. The external circuit is thus switched on or off with a change from light to dark, or vice versa, depending on which of the relay contacts are employed for switching.

The sensitivity of this circuit can be improved by adding a stage of transistor amplification via an OC201 transistor, as shown in Fig. 7.8. Additional components are also included to provide temperature compensation and give a very stable circuit. Component values required are given on the circuit diagram, and again this is a very simple circuit to make up.

CHAPTER 8

PHOTOCELL CIRCUITS AND DEVICES

A PHOTOCELL makes an ideal sensing element for burglar alarms, counters, sorting devices, and smoke detectors, as well as for exposure meters and automatic camera controls. It is widely used commercially for such purposes. The following circuits and devices are primarily designed with simplicity of construction and low cost in mind and all are capable of giving excellent results as practical circuits. Many of the circuits also lend themselves to other applications than the original description.

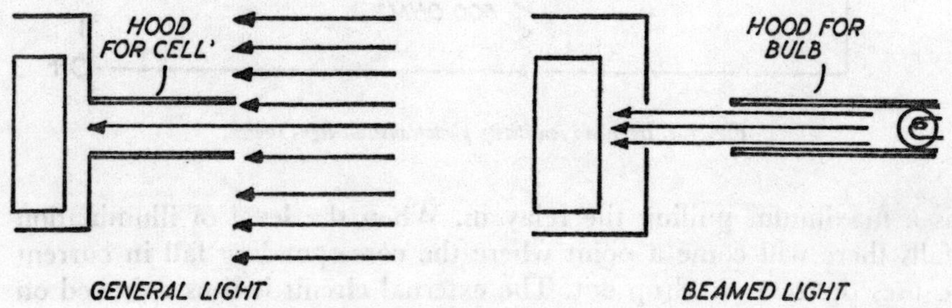

Fig. 8.1. Shielding of photocell by means of a hood (left) and (right) beaming a light source using a hooded bulb.

The source of illumination can be daylight or an artificial light source beamed towards the photocell. In the latter case a simple source of infra-red illumination is a low voltage electric lamp, such as a torch (or simply a torch bulb and battery) with an infra-red photographic filter in front of the bulb. For satisfactory working, or maximum effect, it may be necessary to beam an artificial light source, either with the aid of a simple lens system or merely by enclosing the bulb in a fairly long small diameter tube (see Fig. 8.1). It should also be appreciated that with artificial light sources the level of illumination received by the photocell is inversely proportional to the square of the distance between the light and the photocell—thus halving the distance between light and photocell will increase the light intensity on the photocell by four times. Maximum illumination from any artificial light source is provided when the source is as close as possible to the photocell. Moving the light source farther away will decrease the illumination of the photocell in proportion to the square of the distance.

Provided the photocell itself is adequately shielded and arranged so that the source of illumination is from a positive direction rather than from general illumination, the range which can be achieved is quite considerable. Thus in some cases, using a suitably sensitive circuit, the distance between light source and photocell can be as much as 50 feet or more and still provide satisfactory working. In general, however, where a beamed light source is used it is advisable to locate it as close to the photocell as possible, or convenient. Where the photocell is to operate on the general light level, of course, the question of distance does not come into it.

Simple Burglar Alarm

This circuit uses only a minimum of components and is intended to operate off a 15 volt DC supply. The circuit is shown in Fig. 8.2 and the components required are:

one ORP90 photoconductive cell
one electric bell (intended to operate on about 6 to 12 volts DC)
one two-pole relay with a coil resistance of 2 kilohms (A Post Office
  relay type 3,000 for example, but any two-pole relay with a coil
  resistance of 2 kilohms and capable of pulling in with a current
  of 4 milliamps will do)
one switch or pressbutton
one resistor matched to the bell resistance
source of 15 volts DC (e.g. battery pack).

Wiring connections to the photocell can be made directly to the pins. However, for a permanent assembly it is much better to wire to a B7G valve base when the photocell plugs into this base.

Complete circuit details are shown in Fig. 8.2. This is a very straight-forward assembly and should present no difficulties since the main components—valve base, relay switch and bell—can be mounted directly on to a suitable base panel and then wired up to include the resistor in the circuit.

With the cell illuminated from the correct position (i.e. from the side in the case of the ORP90) and the battery or power supply connected, pressing the reset switch will energize the relay. The circuit will then stay triggered since the relay contacts RL1 close to keep the relay energized when the reset switch is released and the relay contacts RL2 open to break the circuit through the bell.

If now the light source to the photocell is interrupted the resistance of the photocell will rise, causing the current through the relay to

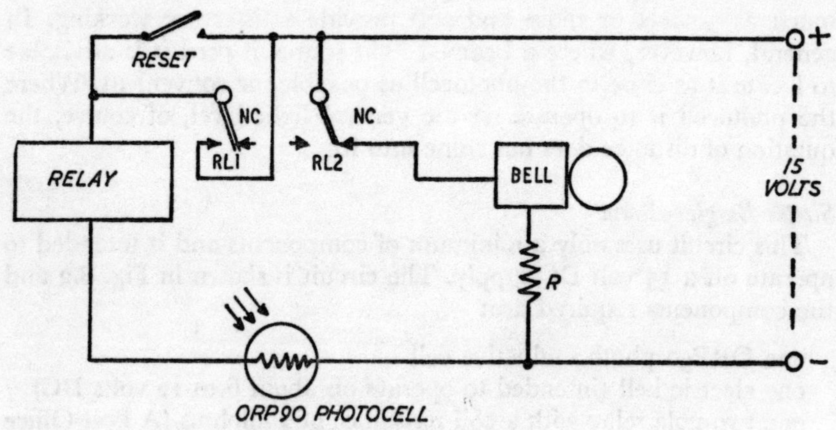

Fig. 8.2. Simple burglar alarm.

fall and the relay to drop out. The resulting closure of RL2 will then complete the bell circuit so that the bell rings and continues ringing even after the light source is restored. The only ways to stop the bell ringing are (i) press the reset button to re-energize the relay and set up the alarm circuit as before; or (ii) disconnect the battery.

There is nothing critical about this circuit (provided a suitable type of relay is used) and the ORP90 photocell will respond either to visible light or infra red. The value of the resistor R in the bell circuit is simply a method of adjusting the current flowing through the bell. If the bell is suitable for 18 volt working (e.g. a 24 volt bell) no resistor R will be required. For a bell designed to work on 9 to 12 volts, the value of resistor R should be about the same as the electrical resistance of the bell coils. For a 4·5 to 6 volt bell the value of R should be about twice the bell coil resistance.

This circuit is capable of being turned into a rather more sophisticated alarm which will ring for a short period only when the light source is momentarily interrupted, then switch the bell off and reset the circuit automatically. The additional components required are a second 2 kilohm relay (RLB), a 1 kilohm resistor and an electrolytic

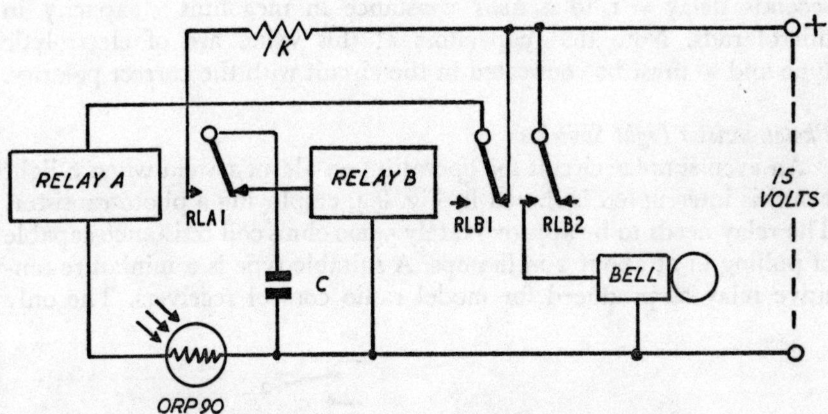

Fig. 8.3. *Modified burglar alarm circuit with automatic circuit reset.*

capacitor of either 1,000 or 2,000 mf value. The modified circuit is then shown in Fig. 8.3. In this case with the light source on, relay A is energized through one set of contacts of relay B, relay B being de-energized. When relay A pulls in its contacts RLA1 provide a patch for charging the capacitor C through the 1 kilohm resistor. When the light beam is interrupted, relay A is de-energized; contact RLA1 changes over and the capacitor is now caused to discharge through relay B, so energizing relay B which pulls in.

The resulting closure of RLB2 contacts completes the bell circuit and the bell will continue to ring as long as the capacitor continues to discharge. When the capacitor has given up all its charge, relay B is de-energized and drops out. This breaks the bell circuit, thus stopping the bell ringing. It also completes the circuit to relay A through RLB1, re-energizing relay A and setting up the circuit once again. The time for which the bell will ring is governed by the value of C and should be about 1 to 2 seconds with a 1,000 mf capacitor and 3 to 4 seconds with a 2,000 mf capacitor.

It is readily possible to adjust this "ringing time" beyond the range quoted, if desired, by using further values of C (or trying different combinations of C and R. As an approximate guide, the delay interval with RC (resistor-capacitor) combinations of this types should be seconds delay = 1 to 2 *times* resistance in megohms × capacity in microfarads. Note that capacitors of this value are of electrolytic type and so must be connected in the circuit with the correct polarity.

Phototransistor Light Switches

An even simpler circuit for operating an alarm system when a light beam is interrupted is shown in Fig. 8.4, employing a phototransistor. The relay needs to be approximately 5,000 ohms coil resistance capable of pulling in at about 2 milliamps. A suitable type is a miniature sensitive relay as produced for model radio control receivers. The only

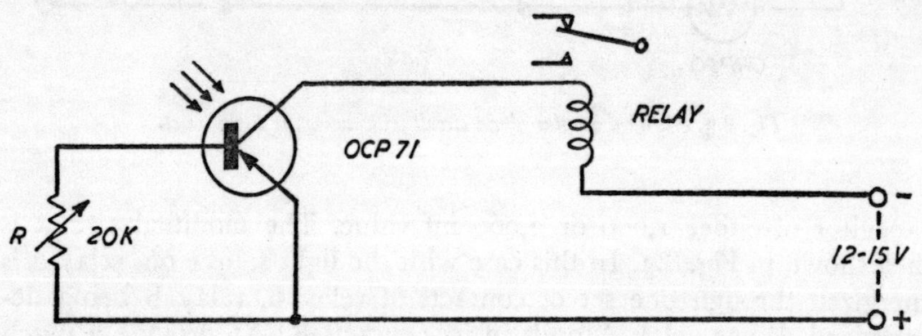

Fig. 8.4. *Simple phototransistor light switch.*

other components involved are a suitable battery or DC supply to match the phototransistor requirements (12 volts) and a 10K or 20K potentiometer. The potentiometer is connected across the base and emitter of the transistor and serves purely as a sensitivity control for adjusting the relay pull-in to the level of illumination available or desired.

No reset function is incorporated since the actual alarm circuit is quite independent of the main circuit with its own separate battery, the circuit being completed by the opening of the relay contacts when the relay drops out at a low level of illumination. This alarm circuit will be switched off again as soon as illumination is restored to the phototransistor, and thus the bell will ring only during the period the light beam is interrupted—just momentarily if anything passes through the light beam. It is thus more correctly described as a photorelay (light-operated relay) or photoswitch circuit.

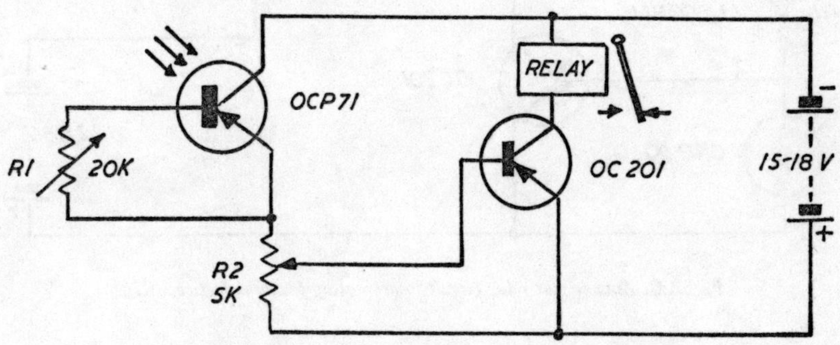

Fig. 8.5. Improved sensitivity phototransistor light switch.

The sensitivity of this circuit can be greatly improved by adding a second (conventional) transistor to provide one stage of amplification following the phototransistor (Fig. 8.5). A second potentiometer (R2) is included to adjust the bias applied to the base of the second transistor which in turn affects the relay current, and thus the pull-in of the relay. In practice, R2 is adjusted with the phototransistor shielded or covered up so that the relay does not pull in with the circuit switched on. The phototransistor is then uncovered and RL adjusted to set the level of illumination at which the relay pulls in.

Annunciator Relay

The circuit shown in Fig. 8.6 employs a conventional photo-conductive cell with straightforward transistor amplification, giving a simple design, again with a minimum of components. Virtually any type of relay can be used with a coil resistance of between 1K and 10K and which can pull in on about 3 milliamps or less. A potentiometer can be used to adjust the relay current to the required operating level. Alternatively, with a relay coil resistance of 5,000 ohms, this potentiometer can be dispensed with. A sensitivity control for the whole circuit is provided by R1, which can be adjusted to establish the pull in

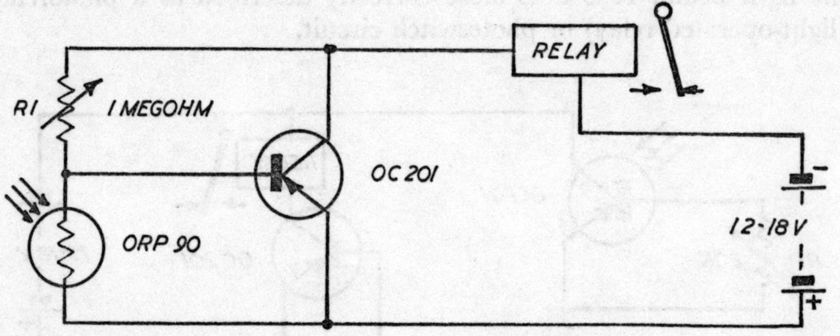

Fig. 8.6. *Annunciator relay circuit incorporating photoconductive cell.*

of the relay at the desired level of illumination. Ideally this should be with as much of R1 left "in" as possible (corresponding, that is, to a fairly high level of illumination) so that the current drawn when the light is on is very low. The circuit can thus be left set with very little drain on the battery. When the light is interrupted or removed from the photocell the current will rise to between 3 and 5 milliamps, causing the relay to pull in and closing the contacts to complete the alarm circuit.

Annunciator Relay with "Hold"

The circuit just described can, with slight modifications, be made to work with a "hold" function. That is, when the relay drops out in response to an interruption of the light it continues to hold out until the circuit is reset.

This is accomplished by completing the circuit through the relay contacts, as shown in Fig. 8.7. In this case only a single battery is required and the circuit is reset by a pushbutton switch. In this respect it is similar to the burglar alarm circuit previously described, except that it used a single pole relay—a relay with a single set of changeover contacts.

The circuit works as follows. With the photocell illuminated, R1 is adjusted so that the relay in not pulled in. When the light level falls and the photocell resistance rises, the balance of the potential divider circuit shifts sufficiently for the current, amplified by the transistor, to operate the relay. The relay contacts now change over, removing the

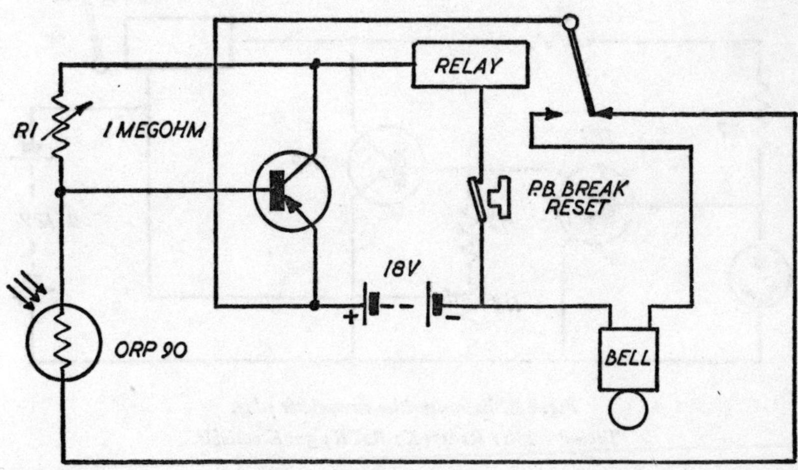

Fig. 8.7. Annunciator relay with "hold".

photocell from the circuit and providing a large positive bias on the base of the transistor. This maintains the current through the relay to keep it held in. At the same time the changeover of the relay contacts completes the alarm circuit. Pressing the reset button de-energizes the relay and restores the photocell to the circuit once more.

Extra-sensitive Annunciator Relay

The addition of a second stage of transistor amplification to the type of circuit already described produces a circuit with extreme sensitivity to light changes. Such a circuit is shown in Fig. 8.8 where the potentiometer R4 controls the overall sensitivity by tapping off a voltage drop which is applied to the base of the second transistor as bias.

A further control can be provided with potentiometer R2 inserted instead of (or in series with) R1. This provides a means of varying the switching time of the circuit or delay between pull-in and drop-out of the relay by adjusting the feedback voltage applied between the collector of the second transistor and the base of the first transistor.

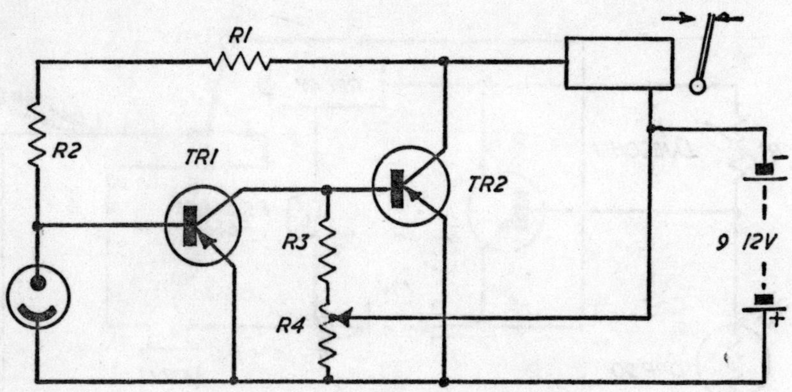

Fig. 8.8. Extra-sensitive annunciator relay.
Typical values: R1 470K; R2, R4 500K variable.

This again is a current rise circuit: an almost negligible current is drawn from the battery when the photocell is strongly illuminated, and the relay is not operated. A fall in illumination produces a change in circuit conditions which triggers the amplifier circuits so that sufficient current is then passed to pull the relay in. The current drain on the battery then rises from a few microamps to the order of 5 milliamps.

External Relay Circuits

All of the circuits described are, in fact, photocell-operated relays, and a relay itself is only an electro-mechanical switch. Where only the relay coil is included in the actual photocell circuit the function of the circuit is simply that of an on/off switch and circuit connected to the switch contacts is quite independent of the photocell circuit. Thus the photocell circuit can be used as a light switch for operating any sort of additional circuit which can be completed by closure of a set of contacts; the closing or actuating of this circuit following from a change of illumination of the photocell.

In addition, this switching action can be used either way, depending on which pair of relay contacts are used as the external circuit switch (see Fig. 8.9).

The switching function provided by photorelays can be used to operate a variety of services by light control, such as opening a door when a light beam is interrupted, operating a counter to record the number of objects or people passing, switch on a light when the level

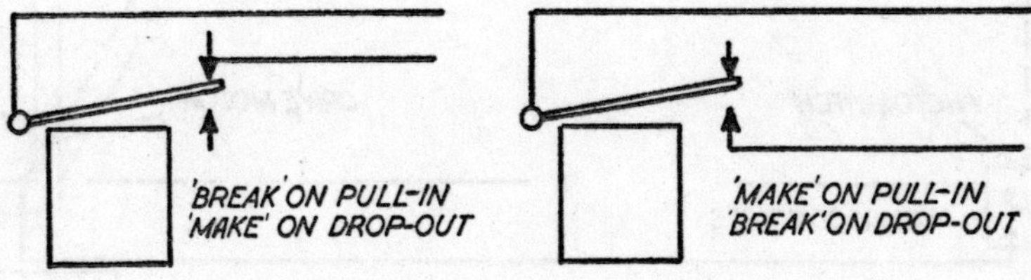

Fig. 8.9. Two different switching actions of relays.

of illumination falls and so on. There are virtually endless possibilities on this theme, but all have the same common characteristic of combining a photorelay with a suitable external circuit. The external circuit can extend a considerable distance from the position of the photorelay, if necessary (see Fig. 8.10, for example, for a diagrammatic representation of a garage door opening system). The appearance of the car interrupting the light beam operates the photorelay, and its contacts then complete the external circuit to the motor which opens the door. The motor circuit would normally be arranged to switch itself off when the door was fully open; or alternatively might incorporate a slipping clutch so that the motor continues to run, but no longer works against the door when fully open. As soon as the car moves forward out of the light beam, restoration of illumination to the photocell then switches the motor circuit off. For closing the door a separate manually operated switching circuit to the motor could be added to drive the motor in the reverse direction. An overriding manually operated switch could also be included in the circuit for the photorelay contacts for opening the door when necessary without interrupting the light beam (or with the photorelay switched off).

In practice, the contacts on the photorelay itself would not be heavy enough to carry the necessary current for such a motor circuit, so it would be necessary to complete the external circuit through a heavy duty slave relay, as shown in Fig. 8.11. The working of such a circuit

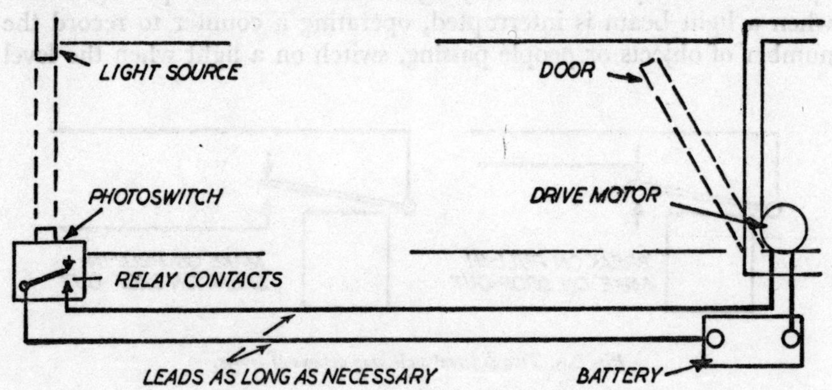

Fig. 8.10. *Basic principles of a garage door opening system using a photoswitch.*

is quite obvious. The (photocell circuit) relay acting as the switch completes the circuit between a heavy duty relay and a battery of suitable size to pull in the slave relay, but not such a large battery that the smaller relay contacts are overloaded. The slave relay contacts then form a further separated switch, and provided they are rated accordingly, can carry currents and voltages well beyond the capacity of the contacts of the first relay. The slave relay, for example, could be selected so that its contacts are rated for carrying a current of several amps at 250 volts or more and could thus act as a switch for a mains supply to an electric motor.

In other cases, of course, the current and voltage in the external circuit may be low enough to be handled safely by the photorelay contacts themselves—for example, to operate a bell in the burglar alarm systems already described, or a simple electromagnetic counter, and so on. The external circuit demand simply sets the requirement for a slave relay, whether this demand exceeds the rating for the original relay contacts or not. Apart from that, the photorelay can be used to act as a switch for *any* type of electrical circuit which works by being switched on or off, whether this circuit is battery or mains operated. In the latter case, however, it is *essential* to separate the photorelay completely from the external circuit via a slave relay.

One limitation of the straightforward photorelay, of course, is that

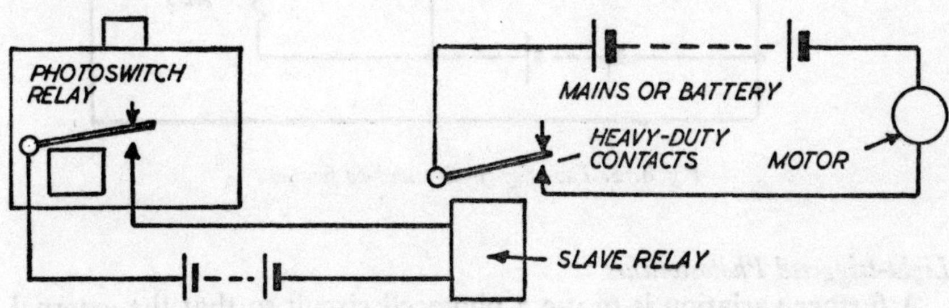

Fig. 8.11. Heavy duty slave relay circuit, this is a practical addition to the circuit in Fig. 8.10.

its switching action is only completed when a light beam is interrupted, unless a "hold" circuit is also incorporated. This means that the light beam must be interrupted for as long as necessary for the working of the external circuit to be completed. Thus the car could not move out of the light beam in Fig. 8.10 without the garage door stopping its opening at that point.

There are many ways of incorporating a "hold" function in the circuit. Fig. 8.12 shows a typical circuit using two relays. This is basically similar to the burglar alarm of Fig. 8.2 with the second relay replacing the bell and the contacts of this second relay forming the switch for the external circuit. When the light beam is interrupted this second relay continues to hold on, and thus hold the external circuit switched on, until the reset button is pressed. There are other methods of applying a similar "hold" function to other photorelay circuits, such as by using a double-pole relay in a circuit which normally does not incorporate the relay switching action in the main circuit.

One set of contacts are then connected to provide a hold function when the relay is operated and the other set form the external circuit contacts.

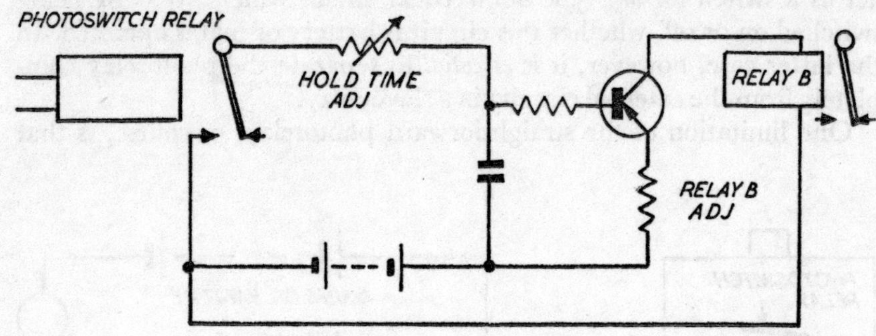

Fig. 8.12. *Two-relay circuit with hold function.*

Light-triggered Photoswitches

A further variation is to use a photocell circuit so that the external circuit is switched on only when the cell is strongly illuminated, and off when the cell is in the dark. This is only a matter of using the opposite pair of relay contacts for the external switch (see Fig. 8.9). The switch is normally shielded from light and its switching action triggered by directing a light beam, such as that from a torch, on to the photocell. This system could be used to switch on a porch light in a house on approaching it in the dark and similar devices of that type.

Again, however, the switching action will only be maintained for as long as the "signal" (or light beam) is directed on to the photocell. The device needs a "hold" function built into it to maintain a switching action. This can be done electronically, on the lines described above, or with a separate bulb circuit, as shown in Fig. 8.13.

Here the relay employed must have two sets of contacts—one set for the external circuit and the other wired to a bulb and battery, as shown. The bulb is placed in front of the photocell so that when lit it illuminates the cell.

When the circuit is originally triggered by the external light beam the relay pulls in and operates both sets of contacts. One completes the external circuit and the other set completes the circuit to light the bulb. Removal of the original light source will then not affect the working since the light bulb circuit will remain complete since the illumination of the photocell is maintained.

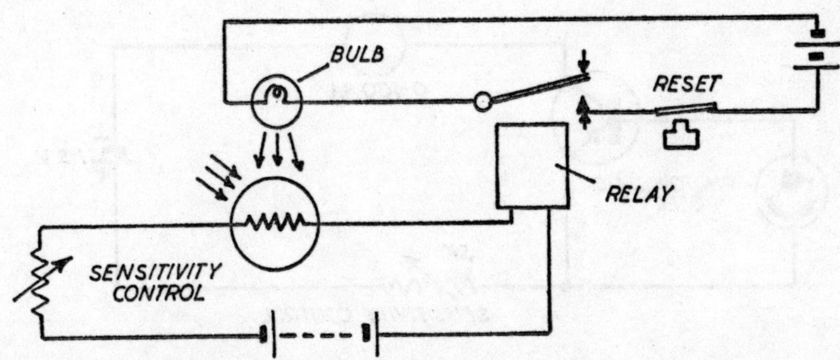

Fig. 8.13. Light-triggered photoswitch with hold function.

A Simple Light Meter

An elementary light meter can be made simply by connecting a photovoltaic cell to a microammeter, as described in Chapter 7. Even with a very sensitive meter, however, meter readings obtained are likely to be crowded into a relatively small movement and a better performance can be realized by applying one stage of transistor amplification to the photovoltaic cell.

A suitable circuit is shown in Fig. 8.14, which also includes a 5K potentiometer as a sensitivity control. A small (1·5 volt) battery is also needed, or a small DEAC cell (1·2 volts). The latter is preferred as being far more stable as regards output voltage than a dry cell and is also rechargeable. Alternatively a non-rechargeable Mallory-mercury battery can be used.

The circuit itself is a very simple one to make, and can readily be fitted into a small box, including the meter, to produce a portable instrument. A switch is necessary in order that the battery can be disconnected from the circuit when the meter is not in use.

The 5K potentiometer or sensitivity control provides a means of calibrating the instrument against either a known light source, or on a comparative basis against an existing light meter. It should be possible, for example, to borrow a photographic exposure meter and calibrate the light meter to read the same under the same conditions. In fact, the scale should be linear as far as light intensity values are concerned. Thus having obtained, say, one "high" light reading by

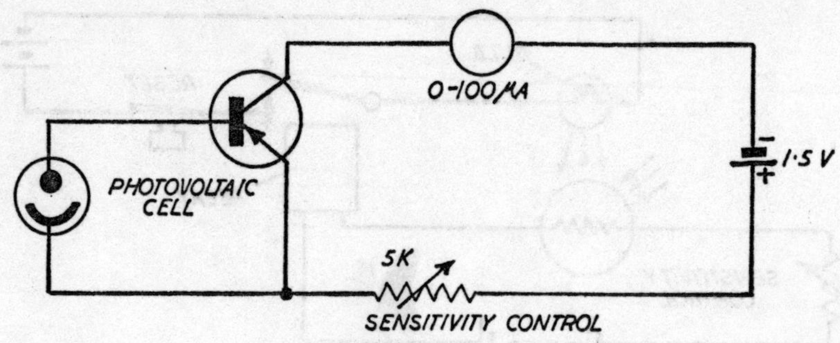

Fig. 8.14. Simple light meter.

comparison, intermediate readings follow by simple division of the scale. Thus the potentiometer needs to be set only once to establish a particular scale figure, scale sub-divisions then following arithmetically as long as the battery voltage remains constant.

Master Photocell Unit

Since photocell circuits, other than photovoltaic cells, have a similar operating principle and function the same circuit or unit can be used for a whole variety of applications. To conclude this section, therefore, details are given of the construction of a professional standard mains-operated unit employing a 90AV photoconductive cell as the light sensitive element combined with a thyratron circuit. Power is supplied via a mains transformer (see Chapter 9).

This is to a design by Mullard and a complete circuit diagram is shown in Fig. 8.15, together with all component values. The whole unit can be housed in a metal or similar box approximately 6″ × 6″ × 3″. Assembly is quite straightforward, utilizing either a conventional metal chassis or breadboard layout, with both the photocell and thyratron being mounted in B7G valve bases. It is desirable to provide a shield around the photocell—a piece of thin aluminium sheet is ideal—and the window in the box should be fitted with a hood roughly 3″ to 4″

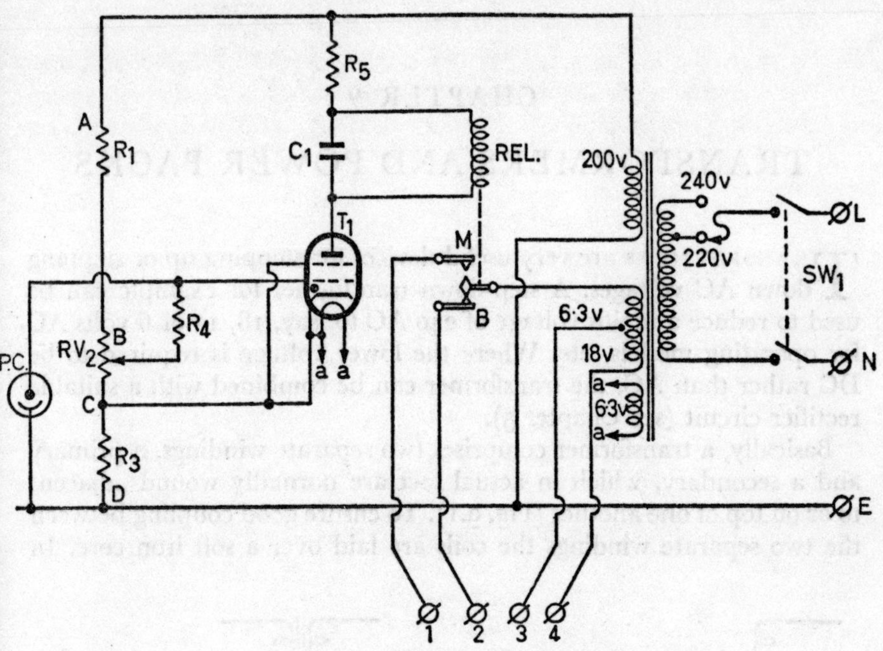

Fig. 8.15. Mullard photocell unit.
Components list:

R1	18kΩ, 2W wire wound	T1	Mullard EN91 thyratron
RV2	5kΩ, 2W wire wound, linear	SW1	Two way, two pole toggle switch
R3	10kΩ, 2W wire wound	Transformer:	Primary 220, 240V, 50c/s
R4	10MΩ, ¼W carbon		Secondary 1-200V
R5	470Ω, 5W wire wound		Secondary 2-6·3V
C1	1μF, 200V working		Secondary 3-18V tapped at 6·3V.
Relay	Post Office type, 2000Ω coil		Also see text.
P.C.	Mullard 90AV or 90CV photocell	Valve holders 2 × B7G.	

long and about 1″ to 1½″ diameter. This ensures that the photocell responds only to light directed into the hood and not to normal background illumination.

The potentiometer (R2) is a sensitivity control for setting up the operating point of the relay. It should be adjusted so that the relay is on the point of operating, but does not actually pull in, when the cell is illuminated. The relay will then operate on quite a small reduction in illumination. This setting corresponds to maximum sensitivity and can be backed off, if necessary, if the unit is to operate over wider changes in level of illumination.

CHAPTER 9

TRANSFORMERS AND POWER PACKS

TRANSFORMERS are very useful devices for stepping up or stepping down AC voltages. A step-down transformer for example can be used to reduce a mains voltage of 240 AC to, say, 18, 12 or 6 volts AC for operating models, etc. Where the lower voltage is required to be DC rather than AC, the transformer can be combined with a suitable rectifier circuit (see Chapter 5).

Basically, a transformer comprises two separate windings, a primary and a secondary, which in actual fact are normally wound adjacent to or on top of one another (Fig. 9.1). To ensure good coupling between the two separate windings the coils are laid over a soft iron core. In

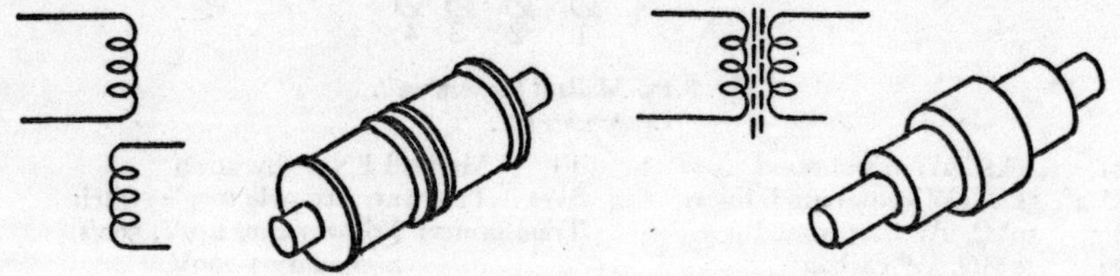

Fig. 9.1. The two basic types of transformer.

the case of power transformers the core, for reasons of magnetic circuit efficiency, normally takes the form of a completely closed circuit of iron with the windings on a central pillar. Also for reasons of efficiency the core is not solid, but consists of thin laminations of soft iron stacked together.

Such laminations are normally available in "E" shape with a matching end "I"—see Fig. 9.2—in ferroxcube E or similar material. Sometimes interleaved "T" and "U" shaped pieces are used to give the same overall form. There is then little more to making a transformer than winding the primary and secondary coils one after the other onto the central bar of the E of a stack of laminations of suitable size and thickness and then completing the transformer by cementing the corresponding stack of "I" pieces in place, using an adhesive such as clear Bostik.

The transformer *ratio* is determined by the ratio of the number of

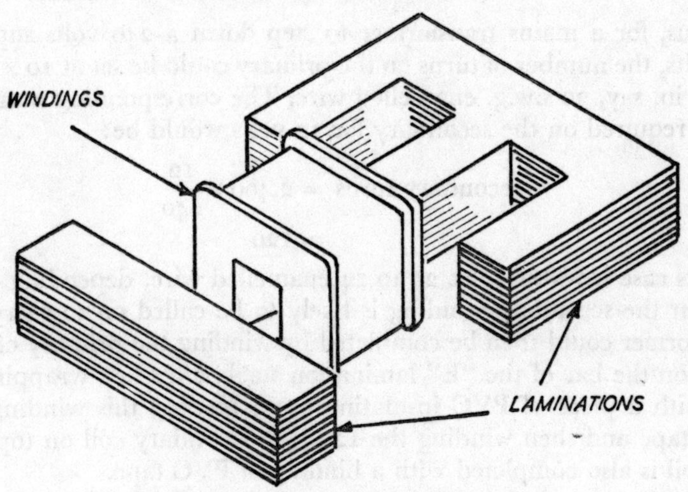

Fig. 9.2. Laminated transformer core.

turns in the primary and secondary windings; providing a step down in voltage if the secondary turns are less than the primary turns, and vice versa. Thus, as a simple rule:

$$\text{No. of turns required on secondary} = \text{No. of turns on primary} \times \frac{VS}{VP}$$

where VP is the voltage applied to the primary
and VS is the voltage required from the secondary.

It is then largely a matter of getting enough turns on to the coils to ensure magnetic saturation of the core and good coupling; and also to use a wire size which can safely carry the current values involved without overheating. A typical design figure for number of turns is 10 turns per volt, and wire sizes as under

mains (primary) windings—26 s.w.g. to 40 s.w.g. (thicker wire sizes usually preferred)

secondary windings—30 or 32 s.w.g. for 200 volts
26 or 28 s.w.g. for 100 volts
20 or 24 s.w.g. for 50 volts
20–26 s.w.g. for 6–12 volts
10 s.w.g. for 1 volt and high currents

Thus, for a mains transformer to step down a 240 volts supply to 12 volts, the number of turns on the primary could be set at $10 \times 240 = 2,400$ in, say, 30 s.w.g. enamelled wire. The corresponding number of turns required on the secondary for 12 volts would be:

$$\text{secondary turns} = 2,400 \times \frac{12}{240}$$
$$= 120$$

In this case we would use 20 to 24 enamelled wire, depending on the current the secondary winding is likely to be called on to carry. The transformer could then be completed by winding the primary of 2,400 turns on the bar of the "E" lamination stack, after first wrapping this bar with a piece of PVC insulating tape; binding this winding with PVC tape and then winding the 120 turn secondary coil on top. This last coil is also completed with a binding of PVC tape.

There are a number of possible variations which can add to the usefulness of a transformer. Thus if the primary winding is "sized" for a certain maximum mains voltage it can be tapped at any number of points corresponding to different input voltages and maintain the same step-down ratio with a fixed secondary coil. Tapping points are calculated in direct proportion (see Fig. 9.3).

Similarly, on the secondary side more than one secondary winding can be employed, each providing a separate output at different voltages according to the number of turns on each coil. Thus a single primary coil can feed two, three or more secondaries supplying different services or circuits (Fig. 9.4).

The following notes give specific instructions for the construction of transformers used with apparatus and circuits described elsewhere in this book. Experience in making one or more of these will be a complete guide for winding transformers with other step-down ratios and as to lamination sizes required, etc. For alternative secondary voltage values it is only necessary to adjust the number of secondary turns on the lines described above.

Alternatively—and this is generally to be preferred for beginners—mains transformers can be bought ready-made in a wide variety of reduction ratios and outputs. This will then ensure a unit which is

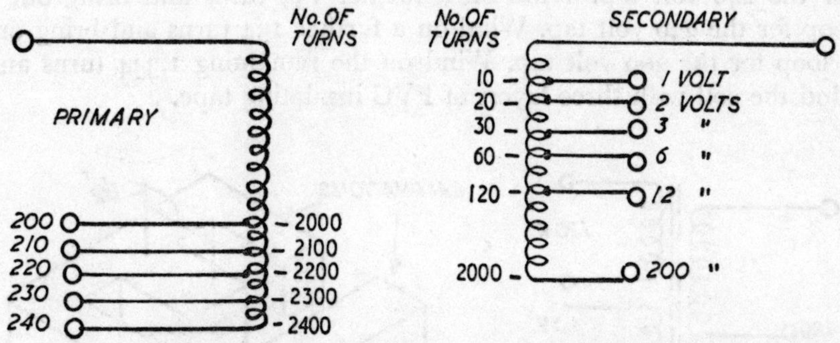

Fig. 9.3. *Tapping points on transformer windings.*

suitable for mains working, and of suitable physical size for a specified output—for example where high currents are going to be drawn from the secondary side the secondary windings must be of generous wire diameter to avoid overheating and possible burning out of the transformer.

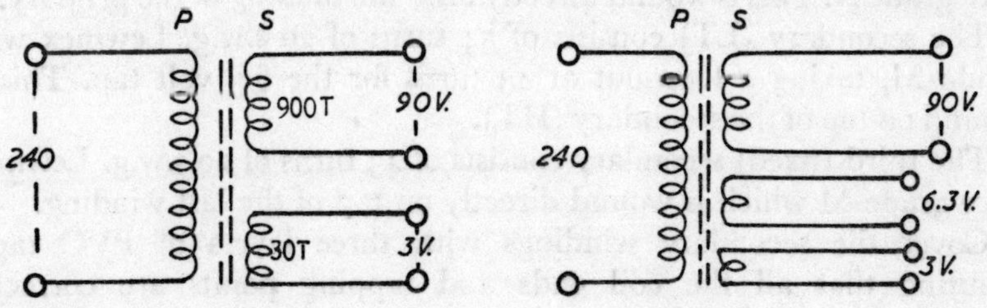

Fig. 9.4. *Diagram showing feeding of two and three secondary windings from one primary winding.*

Transformer for Power Pack

Laminations specified are Pattern 78AN Silcor 25, ·020″ thick. Sufficient laminations are required to make a stack 1″ × 1¼″ (Fig. 9.5).

The primary winding consists of 1,680 turns of 37 s.w.g. Lewmex wire grade F for 250 volts, tapped at 1,612 turns for 240 volts and 1,478 turns for 220 volts and 1,344 turns for 200 volts.

Tape lamination bar and wind on 68 turns, bringing out a loop for the 240 volt tap. Wind on a further 134 turns and bring out a loop for the 220 volt tap. Wind on a further 134 turns and bring out a loop for the 200 volt tap. Wind on the remaining 1,344 turns and bind the coil with three layers of PVC insulating tape.

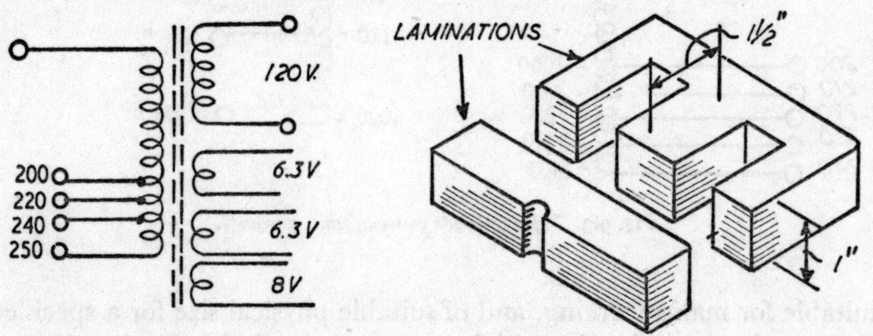

Fig. 9.5. Transformer for power pack. Ratings for primary and secondary windings are shown on the left. On the right are the dimensions of the core stack.

The secondary (HT) consists of 770 turns of 34 s.w.g. Lewmex wire grade N. This is wound directly onto the binding of the primary.

The secondary (LT) consists of 54 turns of 20 s.w.g. Lewmex wire grade M, taking a loop out at 43 turns for the 6·3 volt tap. This is wound on top of the secondary (HT).

The third (fixed) secondary consists of 43 turns of 20 s.w.g. Lewmex wire grade M which is wound directly on top of the last winding.

Cover the secondary windings with three layers of PVC tape, ensuring that all the coil ends and tapping points are correctly identified.

Transformer for Oscilloscope

This is a simple DC converter used in the oscilloscope design described in Chapter 14. Core laminations are ferroxcube E type FX1007 and "I" core FX 1107.

Tape central bar of "E" and wind together two 26 s.w.g. enamelled wires to make 12 complete turns, keeping the wires free from crossovers. Bind with one layer of PVC tape and wind three turns of two wires together on the centre. Cover with one layer of PVC tape and wind on 450 turns of 36 s.w.g. enamelled wire: alternatively (and preferably) wind the 450 turns of 36 s.w.g. wire onto a suitable bobbin and place this bobbin over the central core (see Fig. 9.6).

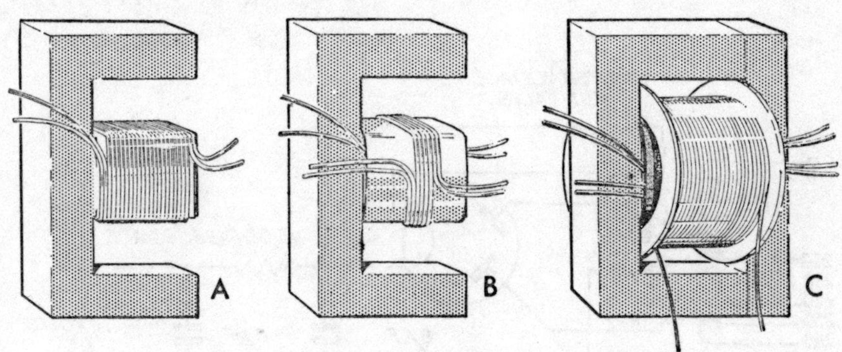

Fig. 9.6. Three stages in making the transformer for the oscilloscope.

Transformer for Master Photocell Unit

This is a transformer for the design described in Chapter 8. In addition to providing all the power requirements for this circuit, additional secondary windings provide a 6 volt and 18 volt output for the operation of bells, lights or alarm or signalling circuits.

The core consists of sufficient "E" stampings and a matching number of "I" stampings to make a stack $1\frac{1}{4}''$ thick—for example MEA Pattern 29A in Silcor 107, $\cdot014''$ thick.

The primary coil wound first on the bar of the "E" comprises 1,440 turns of 40 s.w.g. enamelled and silk covered wire, tapped at 1,320 turns for 200–220 volts.

Wrap with one layer of PVC tape and wind on Secondary 1 consisting of 1,200 turns of 40 s.w.g. enamelled and silk covered wire. Secondary 2 is 40 turns of 25 s.w.g. enamelled wire on top of this; and

Secondary 3 is a further winding of 108 turns of 25 s.w.g. enamelled wire, tapped at 40 turns for 6·3 volts.

Basic Design of Power Pack

The use of a double-diode valve as a full wave rectifier has already been described in Chapter 5 and the circuit of Fig. 9.7 shows a basic 12 volts DC power pack based on such a principle. The mains transformer should incorporate input windings to suit the mains voltage available and secondary windings of 12-0-12 volts (centre tapped), plus

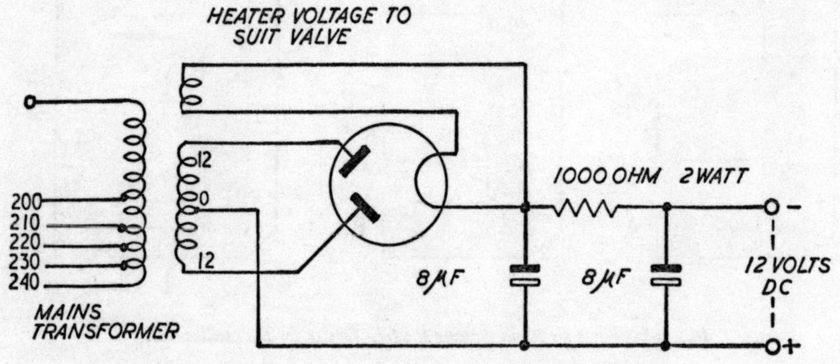

Fig. 9.7. Basic design for 12 volt DC power pack.

a separate low voltage winding to supply the double-diode heater, as shown. Smoothing is accomplished by the use of two 8 microfarad electrolytic condensers and a 1,000 ohm 2 watt rating resistor, connected as shown.

The complete circuit should be mounted on a metal chassis holding the mains transformer (bolted or strapped in position) and the valve base. The whole should then be enclosed within a metal case for safety, with ventilating holes drilled or punched in the side or top of the case. It would also be advisable to include fuses both in the input side (i.e. one lead to the transformer primary); and in the output side (one of the output leads adjacent to the output terminal). The latter will protect the transformer windings against overload in the event of the output being short-circuited.

Rectifiers

Whilst a transformer can step down (or step up) an AC supply, rectification is still essential where a DC output is required, such as that normally used to operate low voltage electric motors, etc. Instead of a diode valve (for large currents) or a crystal diode (for small currents), metal rectifiers can be used since they are simple, compact and trouble-free devices which require no additional current supply, that is no heater current. Thus in the case of the basic power pack described above, a series of four metal rectifiers of suitable rating could be used, connected in a bridge circuit as shown in the first

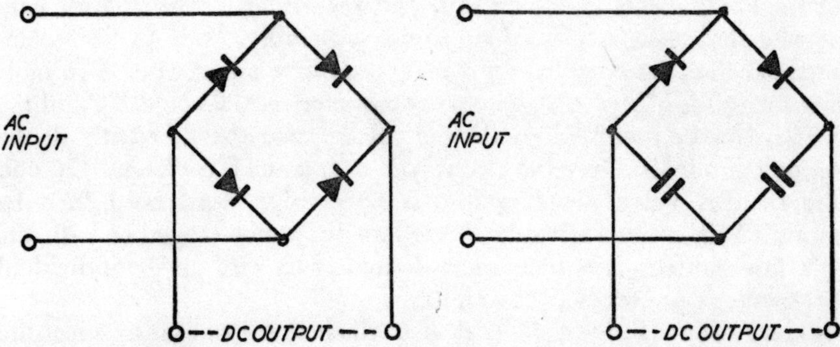

Fig. 9.8. *Two rectifier bridge circuits. Rectifiers have a diode function but require no additional current supply.*

diagram of Fig. 9.8. In this case only a single output is required from the mains transformer, which would involve a single secondary winding with a 17 volt output (for a 12 volt DC supply). The output will be "ripply" but smoothing can be provided, if required, in a similar manner to Fig. 9.7 by incorporating a reservoir and smoothing capacitor in parallel in the DC output circuit, and a 1,000 ohm choke resistor in series in one lead. In some cases the use of the series resistor alone may be adequate for the amount of "smoothing" required.

The second bridge circuit shown in Fig. 9.8 provides full wave rectification with only two rectifiers and with capacitors in the other two legs of the bridge. Provided the output load is high so that the load current is small, this will approximately double the voltage applied from the transformer and also provide some measure of smoothing.

Note that in both these circuits which employ metal rectifiers, the symbols shown are for diodes. Rectifiers have a diode action and so are designated by the same symbol as a diode.

A Complete Power Pack

A complete power pack for providing ripple-free low voltage DC from AC mains must incorporate rectification and smoothing, as well as a step-down transformer. Fig. 9.9 shows the circuit of a suitable design by Mullard, using a double-diode for rectification. Positioning of the components is not critical, and the circuit is suitable for conventional chassis, pegboard or similar assembly. It is to be recommended that the complete unit be housed in a metal case. A suitable size for this is $9'' \times 9'' \times 3''$. The potentiometers R_1, R_2 and R_3 and the switch can be mounted on the face of the case together with all the output terminals. Only the electrolytic capacitors C_1, C_2 and C_3, and the rectifier valve plugging into a B7G valve base, need then be mounted on a separate unit—on a Paxolin panel—together with the two fuse holders. It is then a simple matter to wire up the individual components to complete the circuit.

The cut-out (F_2) is a safety device which operates to disconnect the high tension output in the event of an internal circuit failure. Fuses are incorporated in both the mains input and low tension output as further safety devices.

The unit is intended to cover a wide range of both high tension and low tension voltages likely to be needed for experimental circuits, etc. The high tension DC output can, in fact, be varied from a maximum of 120 volts right down to zero. The two HT outputs (HT 1 and HT 2) differ only in their matter of regulation. HT_1 is varied or adjusted by means of the variable resistor R_1 in series. The output of HT_2 is varied by the potentiometer R_2 circuit.

The other output terminals are arranged to supply low tension up to about 8 volts, the actual voltage being variable by means of R_3. In addition there is a further tapping of the same transformer coil giving 6·3 volts. It will be seen that the low tension output is not rectified and so these are AC voltages. The corresponding maximum currents which can be drawn from the output are 120 milliamps via HT_1 or HT_2,

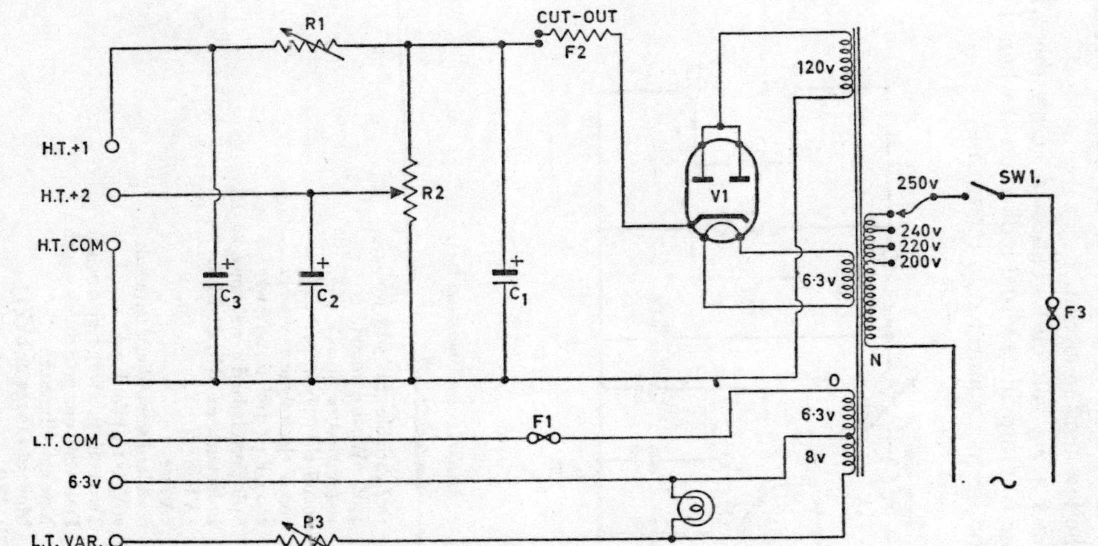

Fig. 9.9. Mullard power pack.

Components list :

V1	Mullard EZ81 with B9A base.		C1	20μF, 250V wkg. Electrolytic capacitor.
R1	5kΩ, 12 Watt, wire wound variable resistor (Colvern Limited).		C2	40μF, 250V wkg. Electrolytic capacitor.
			C3	40μF, 250V wkg. Electrolytic capacitor.
R2	25kΩ, 12 Watt, wire wound variable resistor (Colvern Limited).		F1	5A Cartridge fuse.
			F2	150mA cut-out. (Belling & Lee, type L430).
R3	2Ω, 12 Watt, wire wound variable resistor (Colvern Limited).		F3	2 amp cartridge fuse

and 2·5 amps via the low tension. Any higher currents drawn will cause either the cut-out to operate to break the HT output circuit, or the fuse to blow in the low tension circuit.

The diagram also shows a 2·5 volt bulb connected across the LT outputs. The sole purpose is to provide a visual indication of when the circuit is live (i.e. the transformer connected to the mains).

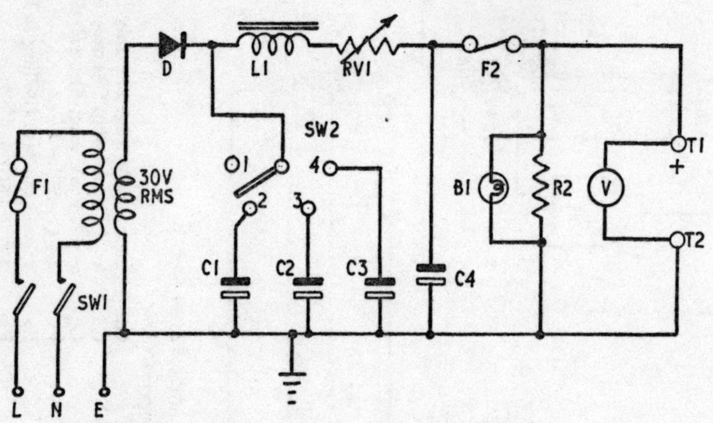

Fig. 9.10. Mullard low voltage power pack.

Components list:

RV1	10Ω variable 40W wire wound
R2	470Ω 2W wire wound
C1	100μF electrolytic 25V wkg.
C2	200μF electrolytic 25V wkg.
C3	400μF electrolytic 25V wkg.
C4	2000μF electrolytic 25V wkg.
L1	Swinging choke d.c. resistance = 5Ω.
	Inductance ≃ 1H., 4A Max.
F1	1A fuse
F2	2A fuse
T1/T2	Standard screw-down terminals
B1	20V, 0·1A pilot lamp
SW1	Two pole, two way toggle switch
SW2	Four way, one pole rotary switch
V	Moving coil meter
DIODE	Mullard OA252 or BYZ13
TRANSFORMER	
	PRIMARY 0–200, 220, 240V a.c. 50 cycles
	SECONDARY 30V r.m.s., 2A.

Simpler Power Pack

A very much simpler circuit—again a Mullard design—is shown in Fig. 9.10. This is a particularly versatile unit in that it can cope with a very wide range of output currents and is thus well suited to the powering of experimental transistor circuits, and so on, in place of dry batteries. It can cope with current demands of from 1 milliamp to 1·4 amps at 10 volts output; and up to 1·9 amps output current at lower voltages, using an input transformer giving a 30 volt (secondary) output.

The specially wound choke coil L1 comprises a winding of 500 turns of 20 s.w.g. enamelled wire on a bobbin with a $1\frac{1}{4}''$ square central hole. This bobbin is then assembled on a stack of "E" laminations $1\frac{1}{4}''$ thick and the unit completed by cementing on a $1\frac{1}{4}''$ stack of "I" laminations with a piece of cartridge paper separating the "I" and "E" laminations to provide a deliberate gap in the flux path. This low resistance choke, in conjunction with the rest of the circuit, provides excellent regulation characteristics.

Switch SW2 has four positions. In position 1 the output of the transformer is fed directly to the choke and the smoothing capacitor C4. In position 2 additional capacity is switched into the circuit, and again at positions 3 and 4. The potentiometer RV1 provides a fine adjustment of the current and voltage supplied to the output terminals, used in conjunction with SW2.

This particular power pack is an extremely useful one for serious experimenters to build. Component positioning is not critical; the only point of major importance being that the silicon diode must be bolted to an adequate heat sink of minimum area 30 sq. cm. and minimum thickness 2 mm. The silicon diode is, of course, the rectifier in the circuit and since only one is employed half wave rectification only is provided, followed by smoothing in the remainder of the circuit. This is the sort of power pack, incidentally, that can be used for operating a transistor radio off mains voltage.

Modern Low Voltage Power Pack

The circuit shown in Fig. 9.11 is a special design for model train and slot-track car operation where the models employ a 12 volt DC electric motor and the continuous current drawn is not likely to exceed 0·5 amps. Where more than one model (motor) is to be operated

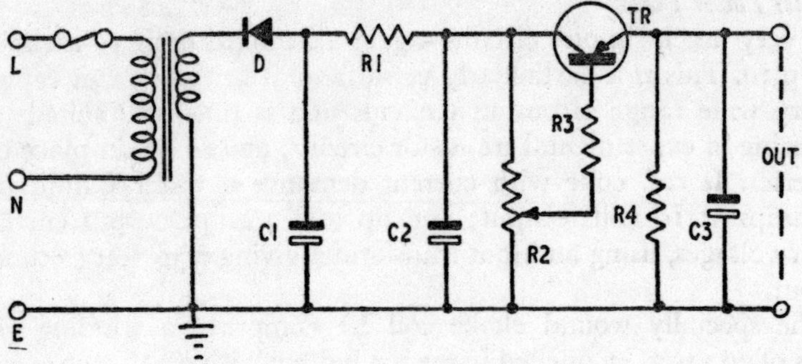

Fig. 9.11. Mullard low voltage DC power supply.

Components list:

Resistors
R1   10Ω   10W
R2   1kΩ   linear potentiometer
R3   120Ω   ½W
R4   120Ω   ½W
Capacitors
C1   2500μF   25V   Mullard C431BB/F2500
C2   2500μF   25V   Mullard C431BB/F2500
C3   0·1μF   125   Mullard C296AA/A100K
            Transistor OC19 or OC25
            Diode AY100

simultaneously, up to three other identical output circuits can be added for separate power supplies.

A 230/12 volt transformer is specified, preferably with double insulation between the windings. A 1 amp fuse is incorporated in the mains (line) lead to the primary. The output from the transformer is rectified by the silicon diode to give a supply which is negative with respect to earth. R1, C1 and C2 act to smooth the DC negative output which is then fed to the collector of an OC 19 or OC 25 transistor. R1 also protects the transistor by limiting the current which can flow through it should the output terminals be short circuited.

The base of the transistor is connected as an emitter follower so that the potential at its emitter nearly equals the potential at its base. The potentiometer R2 provides a means of varying the output voltage from approximately 10v maximum, down to 0v. Resistor R4 provides a path for the leakage current from the transistor when no load is connected to the supply. Capacitor C3 forms a low impedance path for any transient HF currents produced by the motor commutator and prevents them being fed into the supply circuit.

With the circuit shown it is important that neither side of the output be connected to earth as otherwise operation of the reversing switch

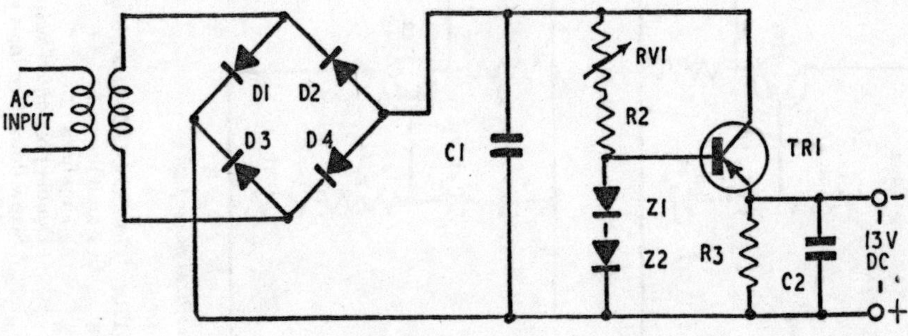

Fig. 9.12. Mullard stabilized 13 volt DC power pack.

could cause a short circuit, reducing the output to zero. The connection of the secondary coil of the transformer to earth via the earth pin of the main lead is a safety precaution.

Both the diode and the transistor used in this circuit require heat sinks to dissipate heat. A suitable heat sink can be cut from 16 s.w.g. aluminium sheet and should not be less than 20 sq. in. in area.

Stabilized 13 Volt DC Power Pack

The circuit shown in Fig. 9.12—also a Mullard design—gives a stabilized DC output; it maintains the output voltage over a wide range of current demands. Normally an increase in current demand tends to pull down the output voltage. Stabilization is achieved by using a suitable transistor (OC29) in an emitter follower circuit with voltage stabilization on the base, provided by zener diodes.

The way in which this works is as follows. Basically the output voltage between the transistor emitter and positive is a function partly of the potential drop across the transistor and partly of the potential difference between the base and emitter. The base-emitter voltage,

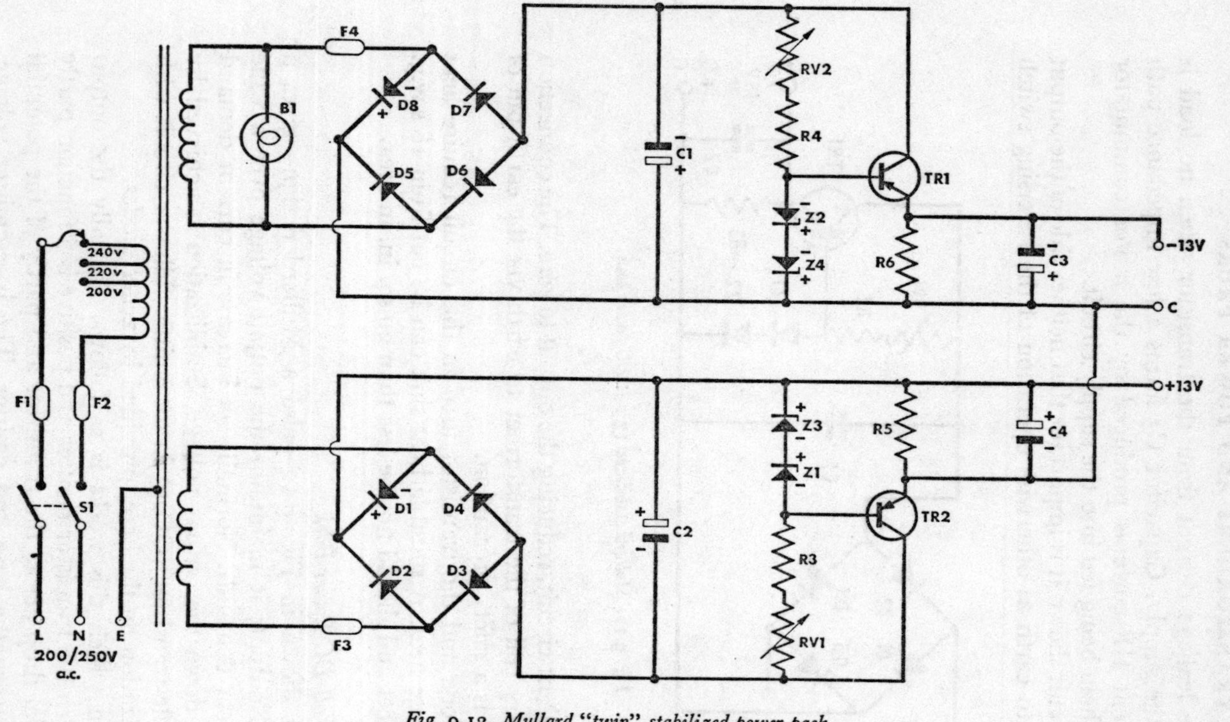

Fig. 9.13. *Mullard "twin" stabilized power pack*

Components list:

RV1, RV2	50Ω, 3 watt pre-set wire wound variable resistor	TR1, TR2	Mullard OC29 transistors
R3, R4	20Ω, 3 watt wire wound resistor	F1, F2	2 amp fuses
R5, R6	100Ω, 5 watt wire wound resistor	F3, F4	3 amp fuses
C1, C2	6,400μF, 25V d.c. electrolytic capacitor	SW1	Double pole on/off switch
C3, C4	2,000μF, 25V d.c. electrolytic capacitor	T1–3	Screw down terminals with 4mm sockets
D1–8	Mullard BYZ13 silicon rectifiers	B1	20V, 0·1A pilot lamp
Z1–4	Mullard OAZ204 zener diodes		

however, is the difference in the output voltage and zener voltage. Thus as zener diode $Z1$ maintains a constant voltage the base-emitter voltage increases as the output voltage decreases; in this respect the circuit is self-compensating. Thus the output voltage remains substantially constant for large changes in output load—for example, for current values from a few milliamps up to about 3 amps.

In constructing this circuit it is essential that the rectifiers, transistor and zener diodes are mounted on separate heat sinks, suitable sizes being:

Each rectifier: 3·75 sq. in. minimum of $\frac{1}{8}$″ thick aluminium
Each transistor: 22 sq. in. $\frac{1}{8}$″ thick aluminium
Each zener diode: 3 sq. in. $\frac{1}{8}$″ aluminium.

The same basic circuit design extended to two outputs is shown in Fig. 9.13. This is capable of supplying two separate outputs of up to 13 volts stabilized DC (actual voltage adjustable via $RV1$ and $RV2$); or 26 volts between the positive and negative terminals of the two output circuits. These are theoretical output voltages corresponding to no current drain. In practice, actual output voltages are 12·75 volts from each circuit used independently; or 25 volts across the two output circuits. Either output circuit used independently can supply a current drain of up to 3 amps. Used simultaneously (either as two separate 12·7v supplies or one 25 volt supply) the maximum total current drain is 4 amps.

CHAPTER 10

AMPLIFIERS, HEARING AIDS AND INTERCOMS

THE function of a multi-element valve in producing multiplication or amplification of a signal has already been described in Chapter 5. We have also seen that transistors also work as amplifiers (Chapter 6). Either valves or transistors, therefore, can be used for a complete amplifier circuit. This can be a circuit on its own—i.e. an amplifier as such—or just a single stage introduced into a more complex circuit to provide amplification as part of the working of that circuit. For example, the signal as detected in a radio receiver is normally so weak that it cannot be heard without first subjecting it to amplification.

A circuit design for a valve amplifier is shown in Fig. 10.1. This can be used on its own to amplify a weak AF input signal to a level where it can readily be heard in a pair of high impedance earphones connected to the output terminals; or amplify a stronger AF signal (one of earphone strength) to a level where it can be heard via a high impedance loudspeaker connected to the output terminals. It can also be used to apply an additional stage of amplification to any experimental circuit involving an RF or AF signal, so it is a very useful circuit to make.

Since a pentode is more effective as an amplifier than a triode this type of valve has been chosen for the circuit. Component values have then been chosen to give stable operation over a wide range. Only the B9A valve base, electrolytic capacitors C_3, C_4 and C_5 and the input and output terminals need mounting on a suitable base plate or chassis. The remainder of the components can be soldered up between the mounted components.

Positioning of the components is not critical, but the input terminals should be mounted close to the valve base so that the wires connecting to the grid components of the valve can be kept to minimum length. This will reduce hum to a minimum, and is a feature which should be followed with all valve amplifiers, incidentally. The valve itself should also be fitted with a screening can when in position. The gain or amplification factor of this particular circuit is approximately 100.

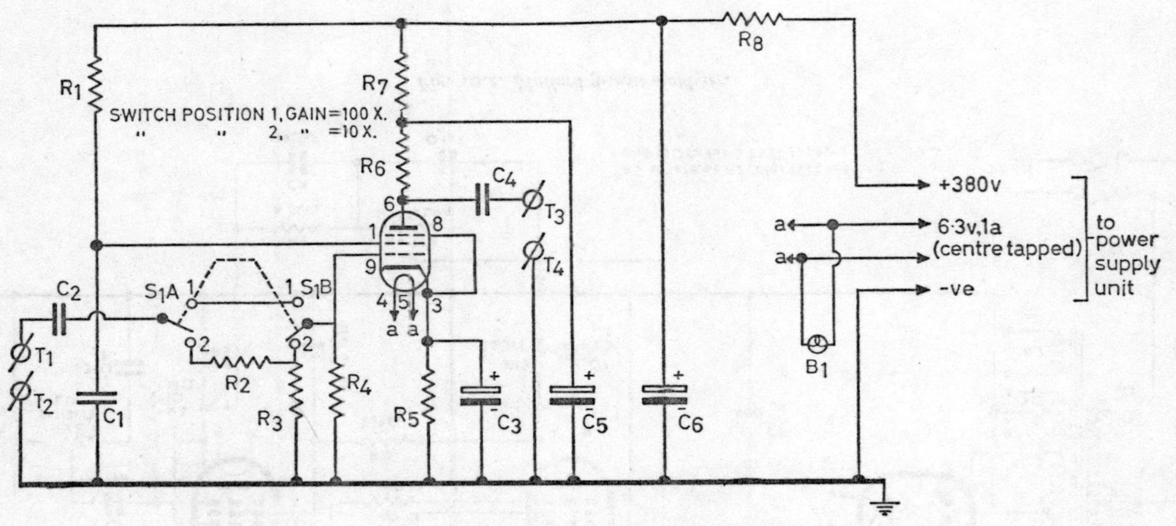

Fig. 10.1. *Mullard high-gain valve amplifier.*

Components list:

R1	390kΩ		C2	0·1μF, paper 350V wkg.
R2	470kΩ		C4	1μF, paper 350V wkg.
R3	47kΩ		C3	500μF, electrolytic 6V wkg.
R4	470kΩ		C5	8μF, electrolytic 350V wkg.
R5	1kΩ	All ¼W 10%	C6	16μF, electrolytic 500V wkg.
R6	100kΩ		V1	Mullard EF86, B9A valve base and screening can
R7	27kΩ		B1	6·3V, 0·2A pilot lamp
R8	39kΩ		S1	Two pole, two way rotary switch
C1	0·1μF, paper 350V wkg.		T1, T2, T3, T4	Standard screw-down terminals

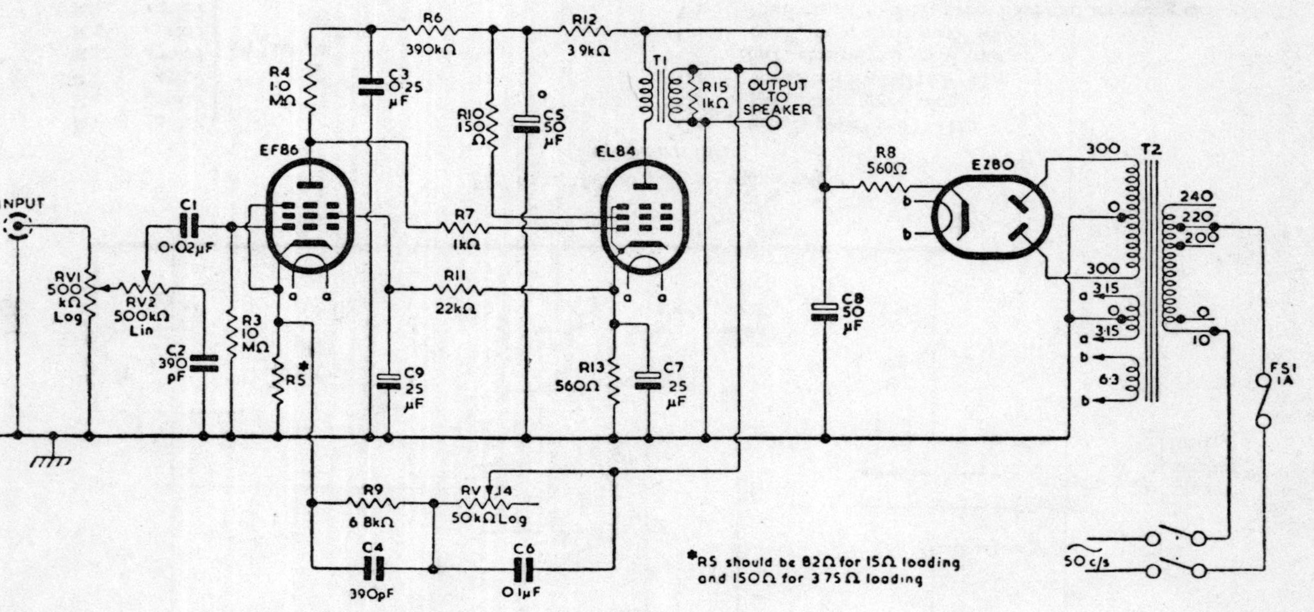

Fig. 10.2. Mullard 3-watt amplifier.

For those wishing to build a more elaborate amplifier, Fig. 10.2 shows the circuit of a 3-valve 3-watt amplifier by Mullard which is capable of providing quality performance at moderate component cost whilst still retaining a basically simple construction. This is capable of handling input signals with a frequency from 35 cycles/second up to 30 kilocycles/second with excellent frequency and power response and also incorporates three controls. Potentiometer RV_1 is a volume control; potentiometer RV_2 is a treble tone control; and potentiometer RV_{14} is a bass tone control.

The mains transformer used should have a rating of 300-0-300V, 60 milliamps with a separate 6·3 volt secondary winding for the filament of the rectifier valve (EZ80). The other 6·3 volt secondary is

---

Fig. 10.2

Components list:

Valves: Mullard EF86, EL84, EZ80.

Resistors

RV_1	500kΩ[1]		
RV_2	500kΩ[2]		
R_3	10MΩ	$\pm$20%	$\frac{1}{4}$W
R_4	1·0MΩ	$\pm$10%	H.S.
R_5	82$\Omega\pm$10% for 15Ω load		$\frac{1}{4}$W
	or 150$\Omega\pm$10% for 3·75Ω load		$\frac{1}{4}$W
R_6	390kΩ	$\pm$10%	$\frac{1}{4}$W
R_7	1kΩ	$\pm$20%	$\frac{1}{4}$W
R_8	560Ω[3]	$\pm$20%	2W
R_9	6·8kΩ	$\pm$10%	$\frac{1}{4}$W
R_{10}	150Ω	$\pm$20%	$\frac{1}{4}$W
R_{11}	22kΩ	$+$10%	$\frac{1}{4}$W
R_{12}	3·9kΩ	$\pm$10%	$\frac{1}{2}$W
R_{13}	560Ω[3]	$\pm$5%	3W
RV_{14}	50kΩ[1]		
R_{15}	1kΩ	$\pm$20%	$\frac{1}{4}$W

[1] Logarithmic; carbon   [2] Linear; carbon.
[3] Wire wound.

Output Transformer T_1
  Primary: 5000Ω.
  Secondary: 3·75Ω or 15Ω.
  The following commercial types have been tested in the circuit and found to be satisfactory:

Manufacturer	*Type No.*
Colne	35206
Gilson	WO767
Parmeko	P2641
Partridge	P4073
Wynall	W.1452

Mains Transformer T_2
  Primary: 10 – 0 – 200 – 220 240V.
  Secondaries: H.T. 300 – 0 – 300V, 60mA.
        L.T. 3·15 – 0 – 3·15V, 1A
            (for EF86, EL84).
            0 – 6·3V, 1A (for EZ80).

If only one 6·3V secondary winding is available, it should have a 2A rating to supply all three valves.

Capacitors

C_1	0·02μF	Paper	150V min.
C_2	390pF$\pm$10%	Silvered Mica or Ceramic	
C_3	0·25μF	Paper	350V wkg.
C_4	390pF$\pm$10%	Silvered Mica or Ceramic	
C_5, C8	50–50μF	Double Electrolytic	350V wkg.
C_6	0·1μF	Paper	150V min.
C_7	25μF	Electrolytic	50V wkg.
C_8	25μF	Electrolytic	50V wkg.
C_9	25μF	Electrolytic	50V wkg.

tapped at the centre to feed the filaments of the two pentode valves. If only one 6·3 volt secondary winding is available on the transformer it should have a 2 amp rating to supply all three valves.

The output transformer (T1) should have a primary resistance of 5,000 ohms and a secondary of either 3·75 or 15 ohms. Suitable commercial types are: Colne 35206; Parmeko P2641; Partridge P4073; and Wynall W.1452.

This particular circuit, incidentally, is also ideally suited for a stereophonic amplifier. In this case two such amplifier circuits are required, one for each channel, although both can be fed from the same power supply. The input stages of both channels are then connected to a three-way selector switch as shown in Fig. 10.3, when the following working is possible:

(i) Stereophonic reproduction from a stereophonic crystal pick-up head.

(ii) Dual-channel monophonic reproduction from a monophonic pick-up head.

(iii) Single channel reproduction (by connecting position b of SW3 to earth instead of position b on SA1

(iv) Dual channel monophonic reproduction from an FM tuner unit

(v) Reproduction of stereophonic transmissions (by connecting position of SW3 to the right hand input terminal instead of the left hand input terminal.

The two amplifier circuits are identical to the single circuit already described (Fig. 10.2) except that RV1 should be a log potentiometer in the left hand channel amplifier and an antilog potentiometer in the right hand channel amplifier. Also the anode ends of the two amplifier circuits are cross connected, as shown in Fig. 10.3. This is a unit for the serious experimenter and one which is capable of giving excellent results.

Transistor amplifier circuits have the advantage of being usually somewhat simpler in circuitry and also demand a smaller voltage power supply which can usually be met by dry batteries. A typical basic amplifier stage has already been shown in Chapter 6. This is the type of transistor amplifier unit normally built into a complete circuit to provide one stage of amplification. It may be followed by one or more similar stages to provide additional amplification, as required. The last stage then provides the output.

A complete transistor amplifier circuit is shown in Fig. 10.4 (page 130). If additional stages of amplification are required then the circuit enclosed within the dashed lines can be repeated. It is as simple as that. The output stage shown used what is known as the half-supply-voltage principle which eliminates the need for an output transformer and minimizes the number of components required in the circuit. The output load is formed directly by a high impedance loudspeaker which drops the supply voltage to one half its value across the speaker.

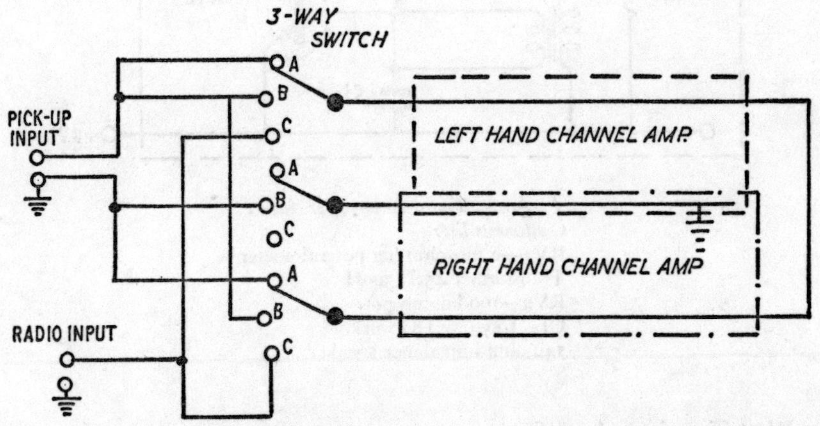

Fig. 10.3. Stereophonic amplifier showing three-way selector switch connections.

The output circuit alone—the RH side of the box—will also work as a complete amplifier, provided the input signal available is not too weak to be heard in the loudspeaker after amplification. It should, for example, be capable of reproducing a good sound level where the input is supplied direct from a crystal pick-up running on a gramophone record.

Where greater output power is required a "driver" output circuit is normally employed. This can take a variety of forms, but the favoured arrangement is usually to employ two output transistors working in a "push-pull" circuit. In such cases, however, it is essential that the pair of transistors used be matched for performance. A typical simple driver circuit is shown in Fig. 10.5. Similar circuits are employed as the output stage in the majority of transistor radio receivers with loudspeaker output.

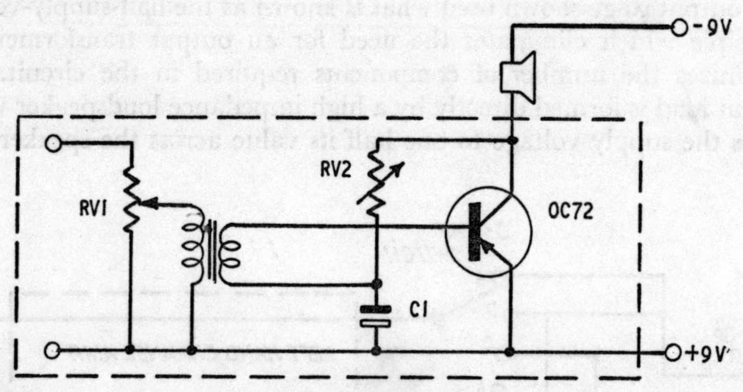

Fig. 10.4. Transistor amplifier circuit.
Components list:
RV1—2 megohm log potentiometer
Transformer 45:1, 500H
RV2—100 kilohm pot.
C1—100μf 3v DC working
140 ohm impedance speaker

5 Watt Transistor Amplifier

A transistor amplifier can be just as powerful, or even more powerful, than a valve amplifier and Fig. 10.6 (page 132) shows a Mullard design for a high quality five-watt amplifier suitable for both mono and stereo applications. It is a simple and economic design which should have a special appeal to the enthusiast seeking quality reproduction from a really compact size.

A circuit diagram only is given, together with a list of components, as this can readily be followed and turned into a practical assembly. Specific details on the construction and performance of this amplifier are, however, also available in leaflet form from Mullards. Printed circuit assembly is suggested, when a panel size of $6\frac{1}{2}''\times3\frac{1}{2}''$ will accommodate all the components together with the mains transformer.

For stereo working two identical amplifiers are required, as already discussed in conjunction with Fig. 10.3.

Hearing Aids and Intercoms

Hearing aids, intercoms and baby alarms are simply amplifier circuits in which the input is provided by a microphone and the output by a loudspeaker. If the speaker can also act as a microphone, and vice versa, the circuit becomes a two-way system or true intercom.

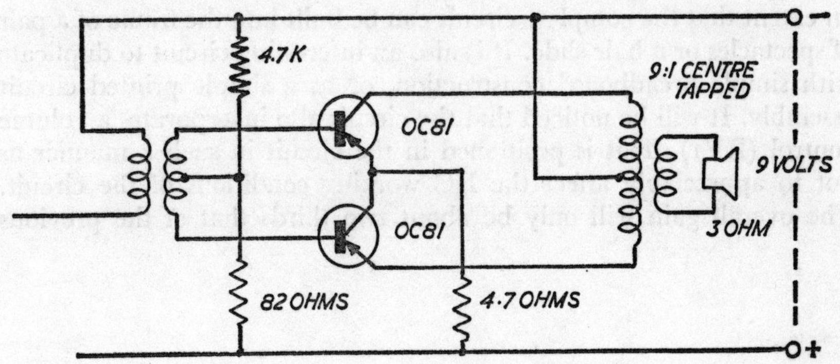

Fig. 10.5. *Simple driver circuit.*

The advent of transistors meant that the bulk of hearing aid sets could be considerably reduced over circuits which had previously to use subminiature valves. The first transistorized circuits used the OC70 or OC71 and one problem associated with the fact that only a very small (2·4 volt) battery was used was stabilization of the circuit at low collector supply voltages. This led to the use of a large number of resistors and capacitors, a typical circuit being shown in Fig. 10.7 (page 134). This is a circuit which can readily be duplicated and has an excellent performance with a low noise level and good frequency response. The overall acoustical quality of the circuit, however, depends primarily on the quality of the microphone and earphone used. Note also that four transistor stages are necessary to produce the required degree of amplification.

With the advent of much smaller transistors specifically developed for hearing aid circuits, it became possible to reduce the number of components required quite considerably, and also the overall size of the set itself. A number of capacitors can be eliminated by employing direct coupling, and it is also possible to reduce the number of transistors to three. This results in about the simplest possible working circuit, as shown in Fig. 10.8 (page 135), which requires only one tiny 1·3 volt Mallory-mercury cell for a battery. The three transistor types OC57, OC58 and OC59 are specially produced for subminiature circuits of this type. All components associated with the circuit are of subminiature type and in a professional set the overall size can be reduced to such an extent that the complete circuit can be built into the frame of a pair of spectacles or a hair slide. It is also an interesting circuit to duplicate with simple breadboard construction, or as a simple printed circuit assembly. It will be noticed that the circuit also incorporates a volume control (RV1). This is positioned in the circuit in such a manner as not to appreciably affect the DC working conditions of the circuit. The overall gain will only be about two thirds that of the previous

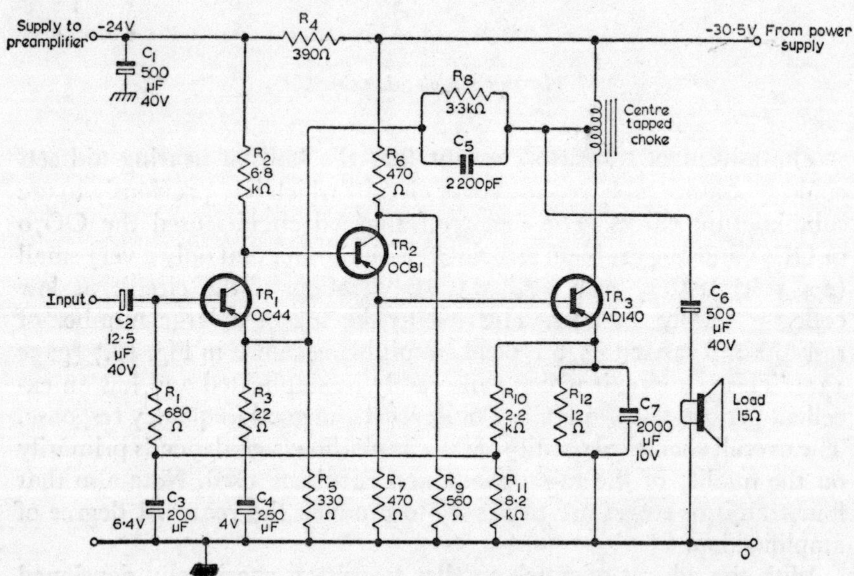

Fig. 10.6. *Mullard 5-watt transistor amplifier.*

circuit described, but the very low current consumption of a little under 3 milliamps gives a life of about 90 hours from a Mallory RM625 cell.

Another modern hearing aid circuit employing three transistors with resistance-capacity coupling is shown in Fig. 10.9 (page 136). This design by Mullard employs the latest hearing-aid transistors and works off a single 1·3 volt Mallory-mercury cell. This circuit has the advantage that it is not critically dependent on the selection of individual transistors as some hearing aid circuits are.

An elementary intercom requires only a very simple circuit comprising a pair of headphone earpiece units in a simple loop circuit

---

Components list:

Transistors
TR1	Mullard OC44
TR2	Mullard OC81
TR3	Mullard AD140

Capacitors   Mullard types

C1	500μF/40V	Electrolytic		C431BR/G500
C2	12·5μF/40V	Min. electrolytic		C426AM/G12·5
or	16μF/40V	Min. electrolytic		C426AR/G16
C3	200μF/6·4V	Min. electrolytic		C426AR/C200
			or	C426AM/C200
C4	250μF/40V	Min. electrolytic		C426AR/B250
			or	C426AM/B250
C5	2200μF/400V	Polyester		C296AC/A2K2
C6	500μF/40V	Electrolytic		C431BR/G500
C7	2000μF/10V	Electrolytic		C431BR/D2000

Resistors

R1	680Ω	10%	1/8W
R2	6·8kΩ	10%	1/8W
R3	22Ω	10%	1/8W
R4	390Ω	10%	1/8W
R5	330Ω	5%	1/8W
R6	470Ω	10%	1/2W
R7	470Ω	10%	1/8W
R8	3·3kΩ	5%	1/4W
R9	560Ω	10%	1/2W
R10	2·2kΩ	5%	1/8W
R11	8·2kΩ	5%	1/8W
R12	12Ω	5%	3W

Choke
Centre-tapped, bifilar wound.
Inductance of full winding, 300mH at 500mA.
Total winding resistance ± 2Ω
Partridge, type TF8000.

with a battery. It will even work without a battery over short distances (see Fig. 10.10 on page 137), but in that case you would not need an intercom for communication at all! However, even with a battery and a longer loop circuit the performance will be poor, reproduction quite distorted and volume marginal. If you are going to make an intercom at all it is worthwhile making a good one.

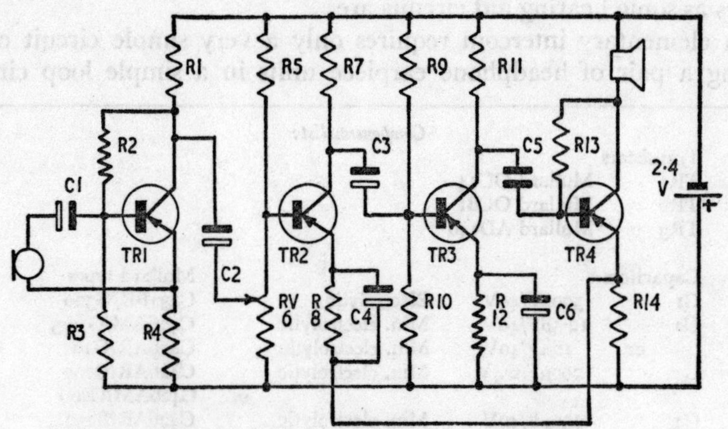

Fig. 10.7. *Older type hearing aid circuit (Mullard).*

Components list:

R1	2·7 kilohm	R13	1 kilohm
R2	56 kilohm	R14	2 kilohm
R3	33 kilohm	C1	8μF
R4	1 kilohm	C2	8μF
R5	18 kilohm	C3	8μF
R6	5 kilohm log pot.	C4	8μF
R7	3·9 kilohm	C5	8μF
R8	1 kilohm	C6	8μF
R9	22 kilohm	TR1	OC70
R10	10 kilohm	TR2	OC70
R11	1·8 kilohm	TR3	OC70
R12	1 kilohm	TR4	OC71

Speaker: 1 kilohm impedance
250 ohm resistance
Microphone: 1 kilohm impedance

All that this really requires is a reasonable amplifier circuit—preferably a transistor amplifier circuit so that it can be worked off a low voltage battery. Identical 3 ohm miniature loudspeakers can then be used both as the microphone and speaker at each station. One station forms the main and incorporates the amplifier. The distant station or extension need then be nothing more than a speaker connected to the main station by a pair of wires. In the case of a baby alarm which requires only one way transmission the extension speaker acts purely as a microphone and is connected by twin flex to the input

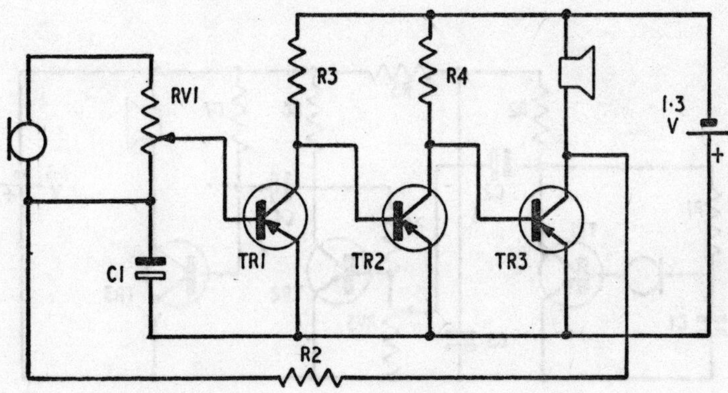

Fig. 10.8. *Sub-miniature hearing aid (Mullard).*

Components list:

RV1	20 kilohm pot.	TR1	OC59
R2	3·9 kilohm	TR2	OC57
R3	selected to match speaker	TR3	OC58
R4	3·9 kilohm		
C1	6μF		
Speaker:	600 ohm impedance	Microphone:	2·5 kilohm impedance
	175 ohm resistance		1 kilohm resistance

of the amplifier at the main station. The second speaker is then connected to the output of the amplifier (Fig. 10.11 on page 137). The amplifier need not be a very elaborate type and a two-stage circuit will probably be quite adequate.

For two-way transmission, or a true intercom, the same system can be used, except that the extension speaker/microphone is connected to a switch at the main station and the main station speaker/microphone also connected to a second bank on the same switch. The action of the switch is to connect the extension speaker to the input and the main speaker to the output for speaking from extension to main;

and reverse the connections so that the main speaker is connected to input and the extension speaker to output for speaking from main to extension (Fig. 10.12).

With a good amplifier circuit this can be a very efficient system with good reproduction and the ability to run the extension wiring up to several hundred feet, if required. By using a switch with additional banks, too, further extensions can be added and selected at will at the

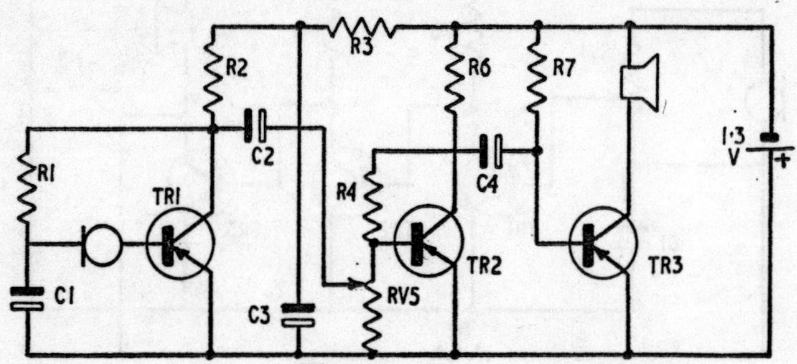

Fig. 10.9. *Three-transistor hearing aid (Mullard).*

Components list:

R1	100 kilohm		C1	2μF
R2	3·3 kilohm		C2	2μF
R3	270 ohm		C3	10μF
R4	100 kilohm		C4	2μF
R5	50 kilohm pot.		TR1	OC59
R6	3·3 kilohm		TR2	OC57
R7	47 kilohm		TR3	OC58
Speaker:	650 ohm impedance		Microphone:	2 kilohm impedance
	200 ohm resistance			

main station to connect with the main station itself. About the only limitation for working is that many transistor amplifier circuits are of a type where the transistors may be overloaded and damaged if the output circuit is open-circuited with the circuit live, that is with the battery switched on. This can be avoided quite simply by using a type of switch for the changeover from send to receive which has a "make before break" action.

A very efficient amplifier circuit specially designed by Mullard for intercom use is shown in Fig. 10.13 (page 138). This employs three transistor stages, the final or output stage being a push-pull circuit to an output transformer. A similar transformer is used for coupling the input to the first stage, and the second amplifier stage is coupled to the driver stage by a third transformer. Suitable types are specified on the circuit

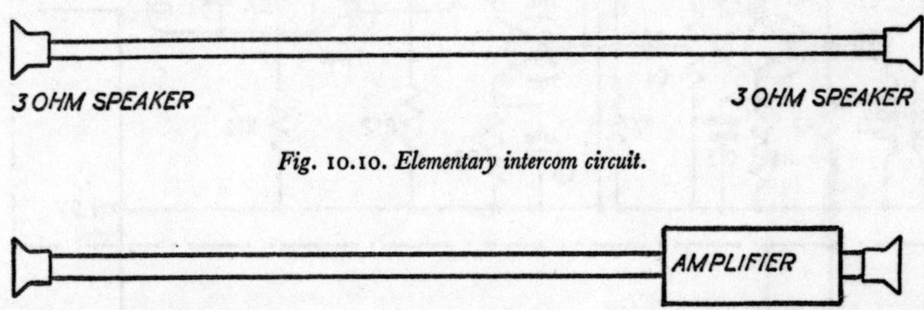

Fig. 10.10. Elementary intercom circuit.

Fig. 10.11. Intercom circuit incorporating amplifier.

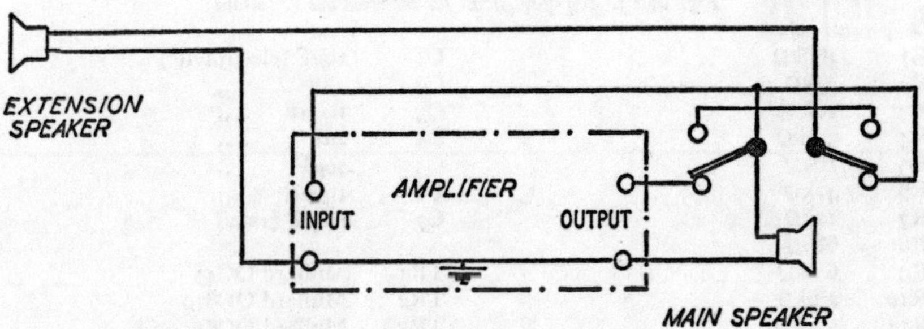

Fig. 10.12. Two-way transmission intercom.

diagram. There is nothing tricky about the circuit or its construction and it needs only a single 9 volt battery—such as a PP3—for power. It is the type of circuit, however, where the switch used must have a "make before break" action and the OC810 and two OC81 transistors will require mounting on heat sinks to ensure that they do not get too hot with continuous working. A suitable heat sink is formed by mounting each transistor on a piece of 16 s.w.g. aluminium sheet 3″ × 2″.

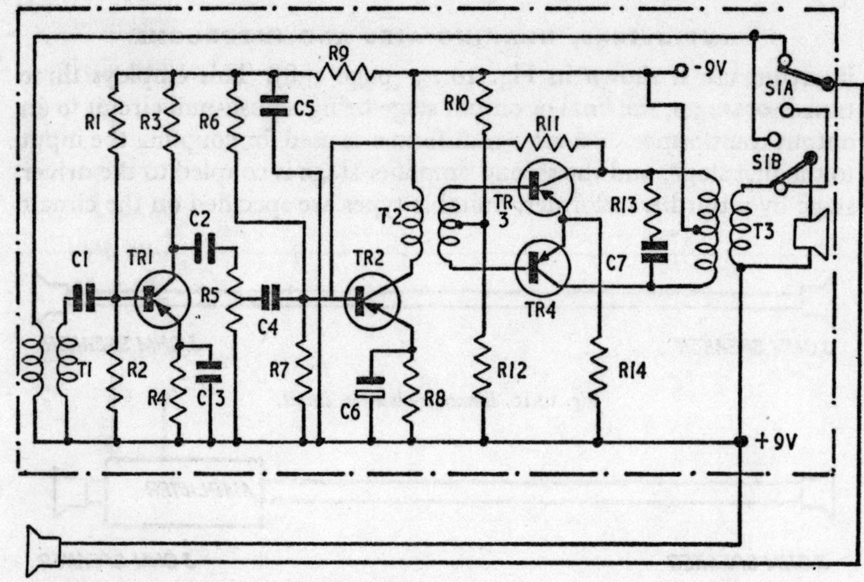

Fig. 10.13. *Amplifier circuit for intercom use (Mullard).*

Components list:

R1	180kΩ	C1	10μF (electrolytic)
R2	33kΩ	C2	2μF	,,
R3	10kΩ	C3	100μF	,,
R4	3·3kΩ	C4	2μF	,,
R5	10kΩ	C5	40μF	,,
R6	47kΩ	C6	100μF	,,
R7	12kΩ	C7	·22μF (125v)
R8	680Ω		
R9	6·8kΩ	TR1	Mullard OC45
R10	2·2kΩ	TR2	Mullard OC810
R11	560kΩ	TR3 }	Mullard OC81
R12	39Ω	TR4 }	(matched pair)
R13	120Ω		
R14	3·3Ω	Loudspeakers: 3 ohm permanent magnet

T1	Ardente D167
T2	Gilson WO929/6
T3	Gilson WO1806

The same basic circuit can be extended to embrace up to three separate extension stations. For example, this can be done by wiring the separate extension speakers to a multi-position switch on the main station for switching in the manner described.

CHAPTER 11

OSCILLATORS

As described in the chapter on amplifiers, the application of feed-back in an amplifier valve can lead to a point where the process of feedback no longer depends on the applied signal and continuous oscillations are generated with the valve supplying its own grid excitation. Thus a valve which is capable of working as an amplifier can also be made to work in a state of self-oscillation where any random variation in current is amplified to cause oscillation. Any circuit capable of generating continuous oscillations is called an oscillator and it will be appreciated that since an ability to amplify is necessary this involves the use of a multi-element valve, such as a triode. A diode does not amplify and so cannot oscillate. A transistor, on the other hand, does have the property of providing amplification and so can also act as an oscillator, in a suitable circuit providing the necessary feedback. The frequency of oscillation in all cases will be that at which the feedback voltage has the proper phase and amplitude.

There is no need to get bogged down too much with theory as basic oscillator circuits are commonly included in parts of complete circuits and thus only need wiring up, with specified components, to perform the function required. We have also seen in Chapter 1 how simple it is to make an oscillator. Some standard forms of oscillator circuits may, however, be mentioned or described by name or type and so for the purpose of completeness we will include brief descriptions.

Two types of oscillator circuits with magnetic feedback are shown in Fig. 11.1; the first being known as a "tickler" circuit and the second as a Hartley oscillator. Both are similar in principle in that feedback is produced by electro-magnetic coupling between the output (anode) and input (grid); in the case of the Hartley oscillator the magnetic coupling is between two sections of the same coil.

Basic circuits with capacity feedback are shown in Fig. 11.2, the first being a Colpitts circuit; the second a tuned-anode, tuned-grid circuit. An even simpler circuit is the crystal oscillator shown in Fig. 11.3 where the crystal itself acts as the tuned circuit in a tuned-anode,

139

tuned-grid oscillator configuration. The other circuit shown is an ultraudion oscillator.

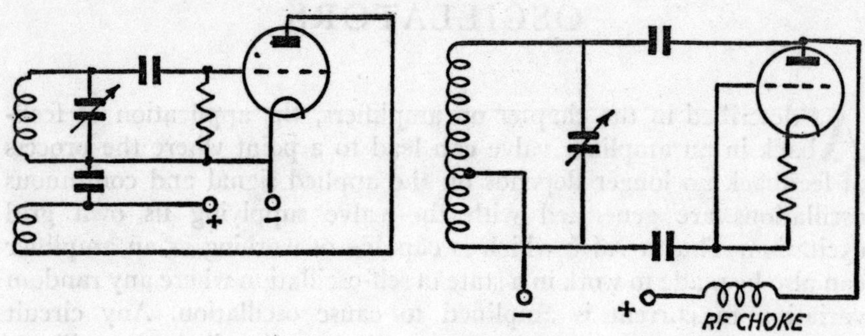

Fig. 11.1. Two types of oscillator circuits with magnetic feedback. On the left is a "tickler" circuit; on the right a Hartley oscillator.

One further type of oscillator worth defining is the transistron—Fig. 11.4. This, in effect, is a sort of perpetual motion device. Theoretically, at least, if a resonant circuit were completely free from losses, a current once started in that circuit would continue to flow

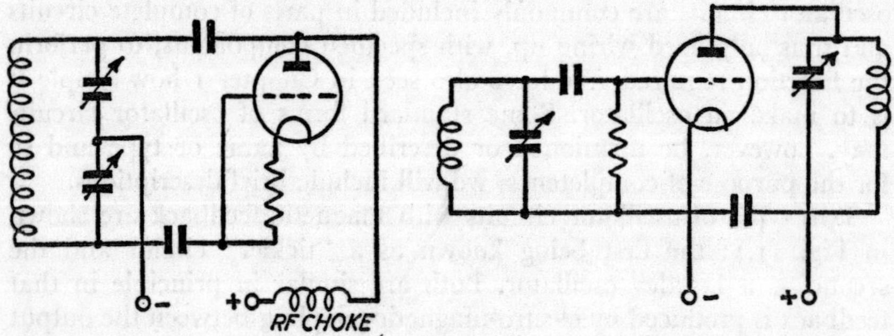

Fig. 11.2. Basic circuits with capacity feedback. Left, a Colpitts circuit; right, a tuned-anode, tuned-grid circuit.

indefinitely. In practice there *are* losses in such a circuit due to real resistances; but the effect of perpetual motion or sustained self-oscillation can be simulated by cancelling the actual resistance with an inserted negative resistance. This means, simply, a device or mode of operation where a decrease in applied voltage is accompanied by an increase in current, and vice versa.

This is the principle of operation of the transitron circuit where "negative resistance" is provided in a pentode valve by virtue of the fact that as the suppressor grid is given more negative bias electrons are turned back to the screen, reversing normal valve action—the principle requirement of the circuit being that the screen grid be

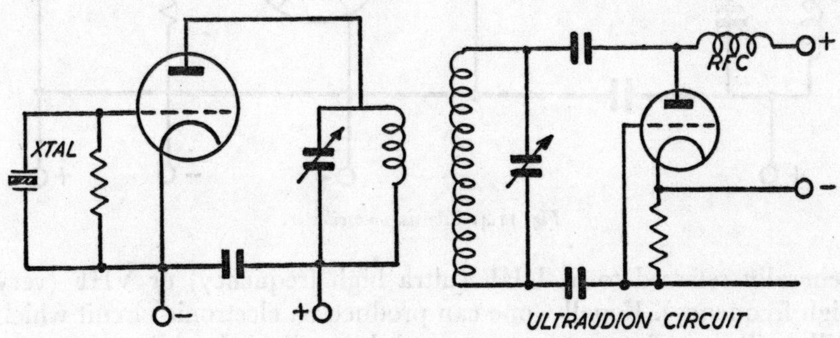

Fig. 11.3. *Crystal oscillator (left). Ultraudion oscillator (right).*

operated at a higher DC potential than the anode. It is not true perpetual motion, of course, since power has to be supplied to the valve to produce this negative resistance effect, but the method is so effective that ordinary tuned circuits based on a pentode can be made to oscillate readily at frequencies up to 15 megacycles per second or more by this simple circuit trickery.

All such circuits, it will be noticed, are grid-leak biased. This is a favoured method since the bias adjusts itself to the excitation voltage available and gives better operation by taking advantage of the grid current flow. All circuits too, have an associated tuned (or tunable) circuit, which is generally referred to as the *tank circuit*. It is this high-Q circuit which determines the frequency of oscillation (i.e. equivalent to the resonant frequency of that circuit). This may be a very high frequency, when the oscillator is an RF (radio frequency) type; or a lower frequency in the AF (audio frequency) range for an AF oscillator.

AF frequency range is roughly from 15 to 15,000 cycles per second; and the RF range from about 10,000 to 3,000,000 cycles per second (10 to 3,000 kilocycles). Radio frequencies above this latter figure are

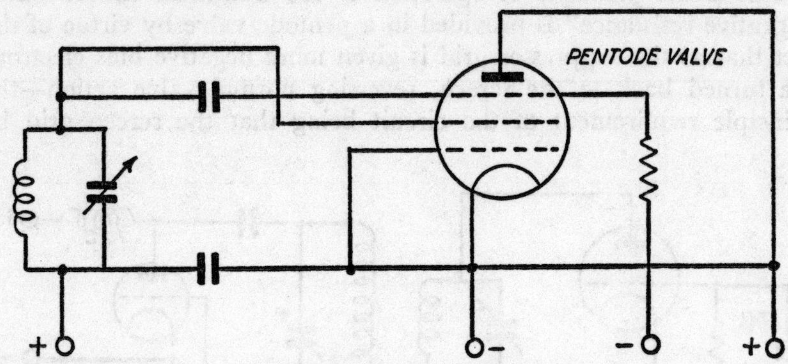

Fig. 11.4. Transistron oscillator.

generally referred to as UHF (ultra high frequency) or VHF (very high frequency). Equally, one can produce an electronic circuit which will oscillate at frequencies very much lower than the AF range.

There is also a simple relationship between *frequency* and *wavelength* which is worth remembering:

$$\text{wavelength} \times \text{frequency} = \text{velocity of light, or}$$

$$\text{wavelength (in metres)} = \frac{300,000,000}{\text{frequency in cycles per second}}$$

$$\text{frequency (in kilocycles)} = \frac{300,000}{\text{wavelength (in metres)}}$$

Signal Generators

One of the most useful applications of an oscillator circuit on its own is as a signal generator or generator of RF or AF waveforms which can be injected into another circuit for test purposes—for example, into an oscilloscope (see Chapter 14), or for aligning a superhet radio receiver circuit.

A simple oscillator circuit based on a single OC72 or OC84 transistor is shown in Fig. 11.5 where the component values are calculated for a frequency of oscillation of 10 kilocycles/second. This is quite a stable circuit and the oscillation frequency can be changed over a fairly wide range by substituting a 0–50 pf variable capacitor in place of the fixed capacitor in the tuned circuit. To perform with complete satisfaction at markedly different frequencies, however, some adjustment of other component values will be found necessary. It is, nevertheless, an interesting circuit to experiment with.

For a signal generator to have maximum usefulness it is necessary to design it to cover a wide range of frequencies, producing a stable

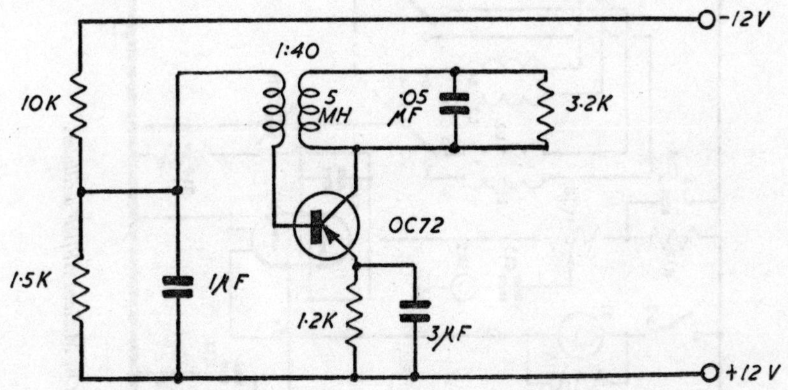

Fig. 11.5. *Simple oscillator circuit for use as signal generator.*

waveform and signal strength and also one which can be calibrated. It will also be necessary to cover both AF and RF, with two separate oscillators, so that the RF can be used to modulate the RF to simulate conventional radio signals (see Chapter 12).

The circuit thus becomes correspondingly more complicated. A typical modern design for an AF/RF oscillator or signal generator is shown in Fig. 11.6. The audio oscillator in this particular circuit is based on using the three transistors in a bridge circuit whilst the RF oscillator section utilizes a low voltage pentode in a conventional transitron circuit. It is the type of circuit specially produced for construction by amateurs and for school use, and is fully described in a Mullard leaflet. It has the virtue of being a battery operated unit

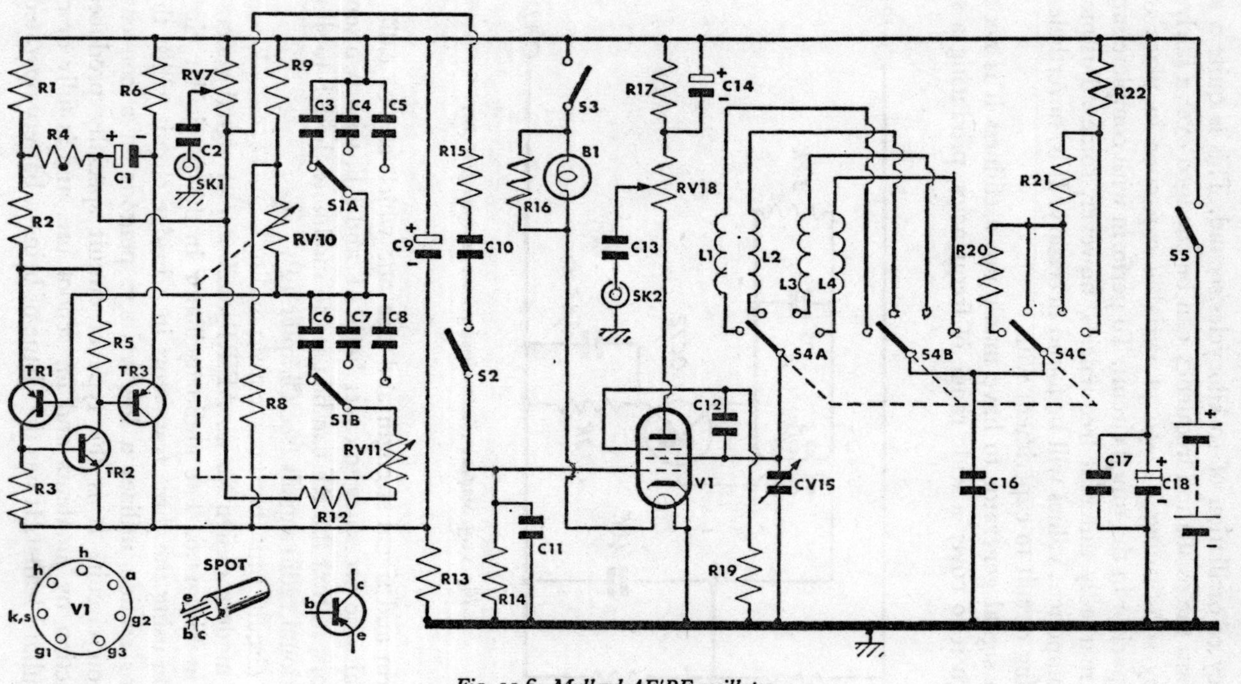

Fig. 11.6. Mullard AF/RF oscillator.

Components list:

R1   100Ω, $\frac{1}{4}$W, 10%
R2   1·2kΩ, $\frac{1}{4}$W, 10%
R3   1·5kΩ, $\frac{1}{4}$W, 10%
R4   STC thermistor type R53
R5   6·8kΩ, $\frac{1}{4}$W, 10%
R6   470Ω, $\frac{1}{4}$W, 10%
RV7   1kΩ, potentiometer, linear
R8   6·8kΩ, $\frac{1}{4}$W, 10%
R9   820Ω, $\frac{1}{4}$W, 10%
RV10   10kΩ, carbon potentiometer ⎫ log. ganged,
RV11   10kΩ, carbon potentiometer ⎭ ±5%
R12   820Ω, $\frac{1}{4}$W, 10%
R13   100Ω, $\frac{1}{4}$W, 20%
R14   220Ω, $\frac{1}{4}$W, 20%
R15   47kΩ, $\frac{1}{4}$W, 20%
R16   47Ω, $\frac{1}{2}$W, 5%
R17   100Ω, $\frac{1}{4}$W, 5%
RV18   500Ω, carbon potentiometer, linear w.w. 5%
R19   1MΩ, $\frac{1}{4}$W, 5%
R20   470Ω, $\frac{1}{4}$W, 2%
R21   1kΩ, $\frac{1}{4}$W, 2%
R22   2·2kΩ, $\frac{1}{4}$W, 2%

C1   1000μF, 15V wkg.
C2   0·1μF, 12V wkg.
C3   0·01μF, 12V wkg., HS close tolerance
L1   440 turns. 42 s.w.g. (0·0040″) enamelled copper wire wound in four groups of 110 turns (see Fig. 4).
L2   300 turns 40 s.w.g. (0·0048″) enamelled copper wire wound in four groups of 75 turns (see Fig. 4).
L3   120 turns 36 s.w.g. (0·0076″) enamelled copper wire, close wound in a single layer.
L4   35 turns 24 s.w.g. (0·022″) enamelled copper wire, close wound in a single layer.

All coils wound on a former, approx. 1$\frac{1}{4}$″ long by $\frac{1}{2}$″ diameter.

C4   0·1μF, 12V wkg., HS close tolerance
C5   4μF, 12V wkg., HS close tolerance
C6   0·01μF, 12V wkg., HS close tolerance
C7   0·1μF, 12V wkg., HS close tolerance
C8   1μF, 12V wkg., HS close tolerance
C9   64μF, 16V wkg.
C10   0·01μF, 12V wkg., 20%
C11   0·02μF, 12V wkg., 20%
C12   0·01μF, 12V wkg., 5% non-inductive mica
C13   0·01μF, 12V wkg., 20%
C14   6·4μF, 6V wkg.
CV15 150pF, tuning capacitor, log. law
C16   0·47μF, 12V wkg., 20%
C17   1000pF, 25V wkg., 20%
C18   500μF, 25V wkg.
S1   two-pole, three-way rotary switch
S2   single-pole, two-way rotary switch

S3   single-pole, two-way toggle switch
S4   three-pole, four-way rotary switch
S5   single-pole, two-way toggle switch
L1   Long wave coil ⎫
L2   Medium wave (1) coil ⎪ see separate
L3   Medium wave (2) coil ⎪ details
L4   Short wave coil ⎭
SK1   Co-axial socket
SK2   Co-axial socket
B1   6·5V, 0·3A, MES pilot lamp and holder
V1   Mullard EF98 and B7G holder
TR1   Mullard OC45
TR2   Mullard OC140
TR3   Mullard OC72
BAT 3×4·5V bell batteries (Ever Ready Type 126, for example)

K

which can be made in a very compact size (say, 12″×6″×3″), requiring only one 12 volt battery.

Light-Operated Oscillator

This circuit, shown in Fig. 11.7, is simplicity itself to construct, and whilst mainly intended as a novelty item could have an application in the form of a warning or monitoring device. It employs a photovoltaic diode or phototransistor to power an OC71 transistor connected as an oscillator and requires no other source of power since no battery is needed to power the circuit. When the photocell is illuminated

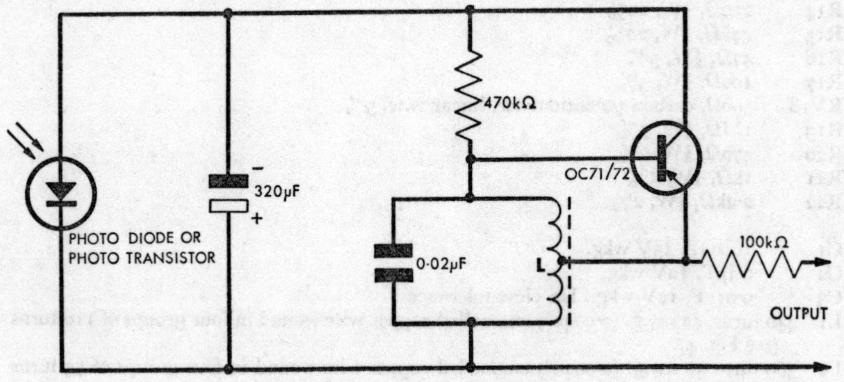

Fig. 11.7. Light operated oscillator.
Components list:
inductance L300+300 turns, 39 s.w.g.
wire on Mullard pot. core FX2238

by a sufficiently strong light sufficient voltage will be generated to operate the oscillator. The resulting output signal from the transistor circuit should be strong enough to be heard in a high impedance crystal earpiece. Alternatively, the output can be amplified by one or more stages added to it, only in this case a battery would be required for powering the amplifier circuit or circuits.

Specification for the coil winding is 600 turns of 39 or 40 s.w.g. enamelled and silk covered wire, wound on a Mullard pot core FX2238 and centre tapped. Resistor and capacitor values are as shown on the circuit diagram. The photocell can be either an OAP12 photo diode or an OCP71 phototransistor. The transistor for the amplifier is an OC71 or OC72.

The Multivibrator

The multivibrator is, basically, a cross-linked circuit between two triodes or transistors both of which are working in an unstable state so that the complete circuit oscillates between the two unstable states. This is initiated by slight unbalance in the components or random variations in current, and so the circuit is self-starting. Typically it

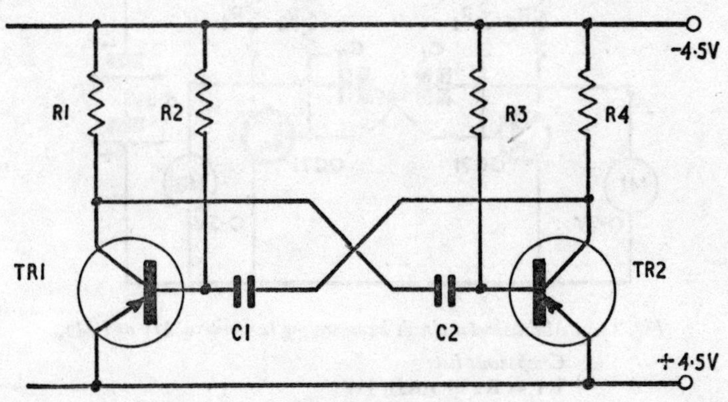

Fig. 11.8. Multivibrator circuit.

Components list:

R1   1 kilohm
R2 and R3 27 kilohm
R4   1 kilohm

C1 and C2 4700 pf
TR1 and TR2 OC71

takes the form shown in Fig. 11.8 where practical component values are inserted to match a pair of OC71 transistors. The frequency of oscillation will be of the order of 5 kilocycles/second with this circuit. A frequency of ten times this value could be obtained from the same circuit using OC41 transistors and reducing R2 and R3 to about 15 kilohms; and C1 and C2 to 1,000 pf.

An extension of this circuit to a very low frequency vibrator is shown in Fig. 11.9 where the circuit actually oscillates at the rate of about one cycle per two seconds. This is sufficiently slow for the two meters incorporated in the circuit to show the operation as the circuit switches from one transistor to the other in turn. Alternatively—and to make a much less expensive demonstration circuit—the meters can be replaced by 3·5 volt torch bulbs. These will then light up and go out alternately, as the multivibrator action switches from one side of the circuit to the other. In order to get sufficient current to light the bulbs, however, two additional OC28 transistors must be included in the circuit for current amplification, as shown in Fig. 11.10.

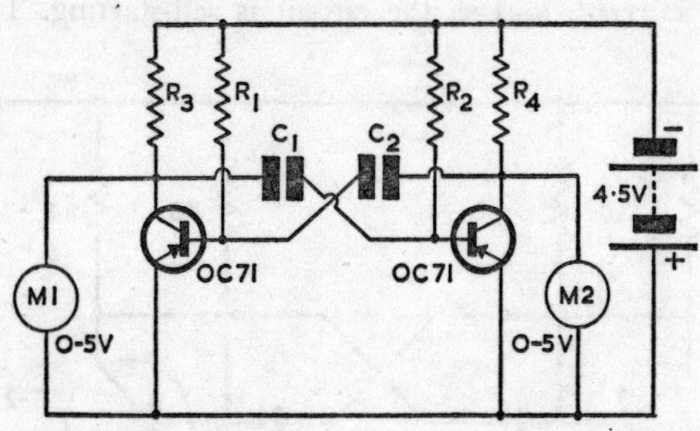

Fig. 11.9. *Multivibrator circuit incorporating two meters, M1 and M2.*

Components list:
$R_1 = R_2 = 10k\Omega$, $\frac{1}{4}$W
$R_3 = R_4 = 1k\Omega$, $\frac{1}{4}$W
$C_1 = C_2 = 100\mu$F electrolytic (6V wkg.)

Making a Metronome

From the basic multi-vibrator circuit to a metronome is a fairly simple step. A suitable circuit is shown in Fig. 11.11 which is wired to provide both visual indication of beats (via the lamps) and audible beats (via the loudspeaker). Furthermore, provision is made to adjust the "beating" frequency from about 22 per minute to 192 per minute via the 250K potentiometer. The OC35 transistor used as a current amplifier can operate up to four lamps without a heat sink. If it is required to operate more lamps than this, then the transistor must be bolted to a suitable heat sink.

The circuit needs no special description since wiring connections are obvious from the circuit diagram and all component values are shown on the diagram. The loudspeaker needs to be of low impedance

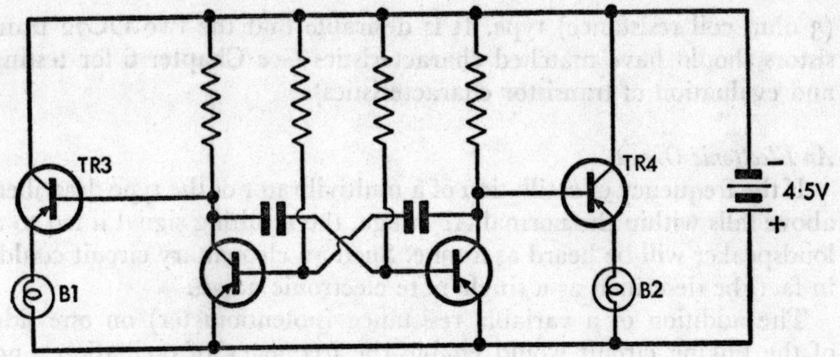

Fig. 11.10. *Multivibrator circuit.*

Components list:

TR3, TR4	Mullard OC28 or similar type
B1, B2	3·5 volt torch bulbs

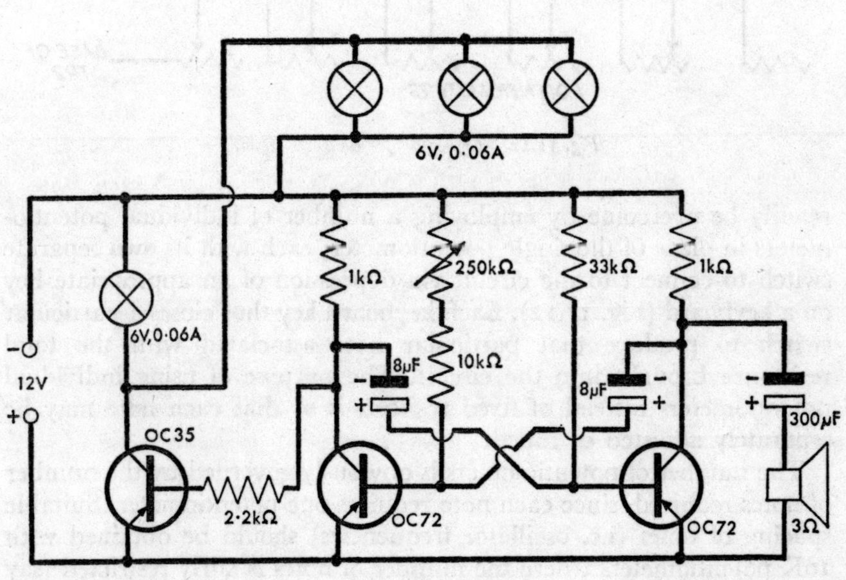

Fig. 11.11. *Metronome circuit.*

(3 ohm coil resistance) type. It is desirable that the two OC72 transistors should have matched characteristics (see Chapter 6 for testing and evaluation of transistor characteristics).

An Electronic Organ

If the frequency of oscillation of a multivibrator of the type described above falls within the normal AF range, the resulting signal if fed to a loudspeaker will be heard as a tone. Such an elementary circuit could, in fact, be described as a single note electronic organ.

The addition of a variable resistance (potentiometer) on one side of the linking circuit would enable the frequency of oscillation, and thus the tone, to be varied, although this would not be a practical way of playing the elementary organ. However, the difficulty can

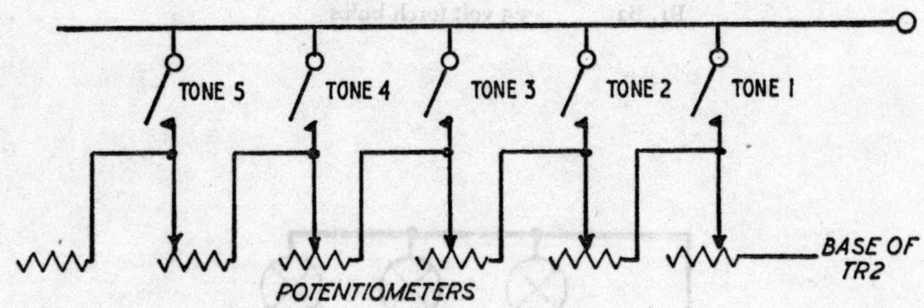

Fig. 11.12. "Keyboard" for electronic organ.

readily be overcome by employing a number of individual potentiometers in place of the single potentiometer, each with its own separate switch to connect to the circuit via depression of an appropriate key on a keyboard (Fig. 11.12). Each keyboard key thus closes a particular switch to produce that particular note associated with the total resistance brought into the circuit. The purpose of using individual potentiometers instead of fixed resistors is so that each note may be separately adjusted or tuned.

The number of potentiometers is obviously governed by the number of notes required, since each note requires one potentiometer. Suitable spacing of tones (i.e. oscillator frequencies) should be obtained with 10K potentiometers where the number of notes is fairly restricted (say not more than 16), otherwise 5K potentiometers would be advisable.

A complete circuit diagram for a simple electronic organ of this type devised by Mullard is shown in Fig. 11.13. This incorporates additional features to the multivibrator circuit—notably a tremolo circuit added to enrich the sound and an amplifier. The tremolo circuit uses an $OC75$ transistor as an oscillator with its frequency governed by R_3, R_4, R_5, C_1, C_2 and C_3. The tremolo output is fed via R_7 and switch SW_1 to the base of either TR_2 or TR_3. The switch is incorporated to switch out the tremolo when not required. Alternatively, of course, the complete tremolo circuit can be omitted as not necessary for the working of the organ.

The multivibrator (tone generator) output is taken from the emitter of TR_3 via a coupling capacitor C_7 to the base of TR_4. The collector load of TR_4 is a high impedance loudspeaker across which half the supply voltage is developed. This produces a stable output circuit with a minimum of components. Power is supplied by a single 9 volt battery and the output is of the order of 50 milliwatts which is quite adequate volume for an average size room.

The circuit itself is quite a simple one to build, and thus as far as the electronics side is concerned this can be made by anyone with little or no previous experience. The making of a suitable keyboard is a mechanical job which may be a little tricky. This is mainly because of the number of individual keys required to get full range of notes for satisfactory organ coverage. Basically, however, each key is nothing more than a simple electrical switch which closes its particular potentiometer circuit when depressed. It could, in fact, be nothing more than a spring brass strip firmly anchored at one end and touching a contact when depressed by the finger.

Logically, two keyboards should be built for the organ—one for treble notes and one for base notes. Each keyboard will then require its own note generator circuit and amplifier, and tremolo circuit (if required). That is, the circuit shown overleaf (Fig. 11.13) should be built twice— one circuit for coupling to the treble keyboard and one circuit for coupling to the base keyboard. The power supply can still be a single 9 volt battery. Component values are the same in each case except for the capacitors C_5 and C_6—see Components List.

In its simplest form the electronic organ can be made as a single keyboard unit coupled to a single tone generator/amplifier circuit, omitting the tremolo circuit.

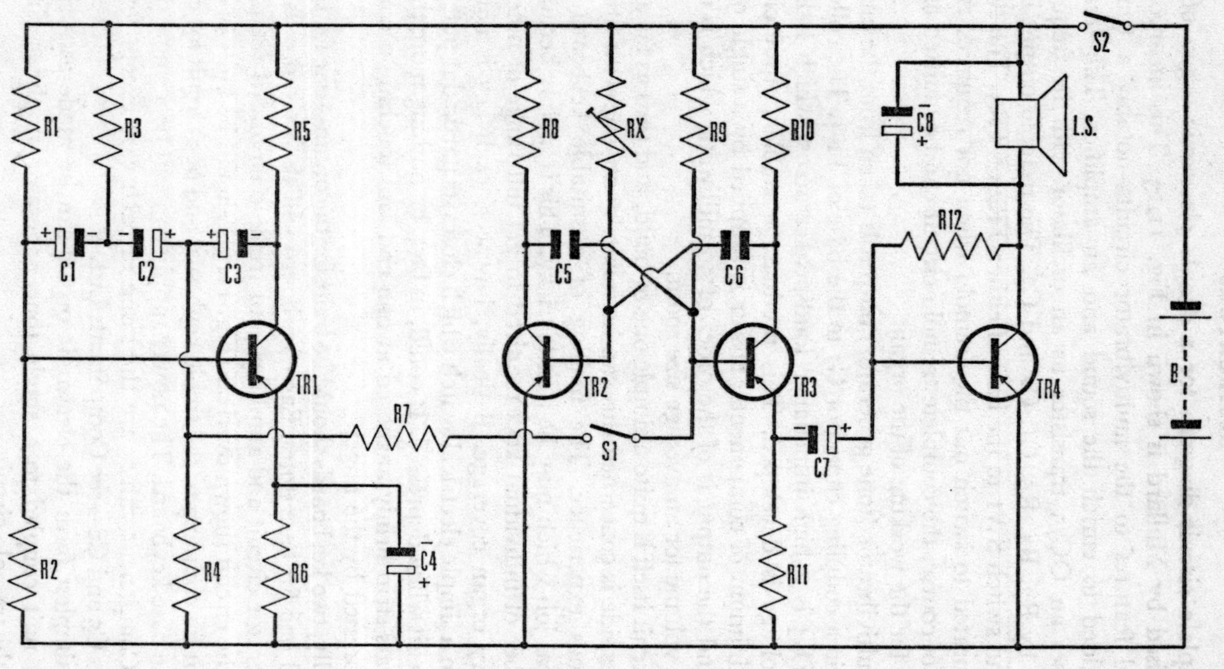

Fig. 11.13. Electronic organ.

Components list:
The following components are for one half of the organ less the preset potentiometers attached to the keyboard.

R1 82kΩ
R2 10kΩ
R3 4·7kΩ
R4 4·7kΩ
R5 4·7kΩ
R6 1kΩ
R7 100kΩ
R8 2·2kΩ
R9 12kΩ
R10 2·2kΩ
R11 100Ω
R12 39kΩ
Rx series of potentiometers (see text)
All resistors ¼W, 10%
C1 2μF electrolytic
C2 2μF electrolytic
C3 2μF electrolytic
C4 250μF electrolytic
C5 0·1μF (treble circuit)
C5A 0·25μF (base circuit)
C6 0·1μF (treble circuit)
C6A 0·25μF (base circuit)
C7 2μF electrolytic
C8 2μF electrolytic
All capacitors 12V wkg. minimum
TR1 Mullard OC75   S1, S2 (single-pole, two-way toggle switches)
TR2 Mullard OC71
TR3 Mullard OC71
TR4 Mullard OC72   B   9V battery (Ever Ready type PP9 or similar)
LS.   80Ω permanent magnet loudspeaker

CHAPTER 12

RADIO RECEIVERS

THERE is nothing at all mysterious about radio receivers. They work on well established electronic principles which are readily understood with a little study. Furthermore the advent of the transistor has made it possible to produce working receivers with a minimum of components and very simple circuitry. A further advantage, too, is that transistor receiver circuits need only a single low-voltage battery, dispensing with the high tension and low tension batteries required for valve receivers. Alternatively they can use a mains supply via a power pack comprising a transformer-rectifier circuit. Descriptions of receivers will, therefore, be confined to transistor circuits.

A typical radio signal as sent out by a transmitting station comprises a high frequency or RF signal known as the "carrier", on which is

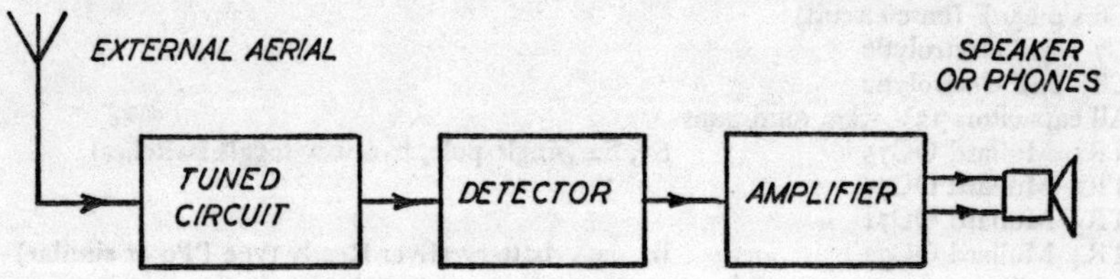

Fig. 12.1. Circuit stages of radio receiver.

superimposed a lower frequency (AF) "tone" signal. This is a signal in the audio frequency range, or frequency range which corresponds to audible sound. The purpose of any receiver is to pick up the modulated radio frequency signal (i.e. the carrier RF on which the AF or tone signal is superimposed), and from it extract the original audio frequency signal. The strength of this AF signal is then boosted, as necessary, by amplification up to a level where it can operate headphones or a loudspeaker and so be heard.

The circuit stages necessary to do this comprise an aerial connected to a tuned circuit, followed by a detector stage, and then one or more stages of amplification (see Fig. 12.1). The function of the *aerial* is to collect any RF signals which may be present at that point, regardless of frequency. Thus all transmitted signals with a range to reach the aerial will, in fact, induce RF signals in the aerial. These signals will be microscopically small in strength, but they will be there.

The purpose of the tuned circuit which follows the aerial is to select *only* that signal frequency required—i.e. the particular signal or station to which the receiver is to be tuned. It does that by virtue of the fact that it is a resonant circuit which responds to, and amplifies, just one particular signal frequency. The tuned circuit thus provides selection of signal and a certain degree of amplification. This selected signal is then passed to the detector which is, in effect, a rectifier which passes only the AF component from the incoming signal in the form of a varying DC current. This AF signal is then passed on to the amplifier for boosting in strength, as necessary, and finally to the loudspeaker.

Having understood this, we can now describe the various stages in more detail. The aerial we need not bother with unduly, except to note that for optimum efficiency its length should be equal to or a multiple or fraction of the wavelength of the signal. This is obviously impractical where a receiver is required to tune into various stations with different signal wavelengths. The usual way of getting around this is to design an efficient tuned circuit which can dispense with an actual aerial wire attached to it. However, it may still be necessary or desirable to add an aerial wire to improve reception, especially in areas of low signal strength—areas of poor reception, either a long distance from transmitters or blanketed by natural conditions.

A tuned circuit consists essentially of a coil or inductance and a capacitor, normally connected in parallel (Fig. 12.2). This combination of inductance and capacity will have a natural or resonant frequency. To tune the circuit to a particular radio frequency, therefore, the values have to be selected so that the circuit has that particular resonant frequency.

The resonant frequency of an inductance-capacitor combination can be calculated from the formula

$$\text{resonant frequency} = \frac{1}{2\pi \sqrt{LC}}$$

where L is the coil inductance and C is the capacitor capacity

or   resonant frequency $= \dfrac{\cdot 1593}{\sqrt{LC}}$
(cycles per second)

where L = inductance in henries
C = capacity in farads

In practice the resonant frequency needs to be adjustable and this can readily be achieved by making one or other of the two components variable. Thus the circuit can be tuned by variable capacity (a variable condenser); or by variable inductance. In the latter case this means winding the coil on a suitable former which is fitted with an iron-dust

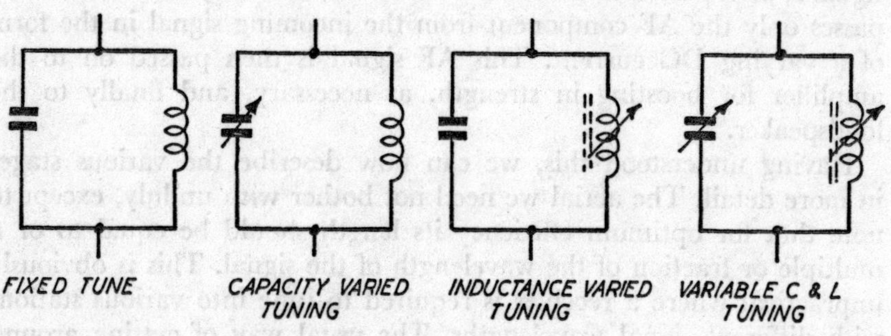

Fig. 12.2. *Types of tuned circuit for radio receivers.*

core. Screwing the core in or out of the former then varies the inductance of the coil itself. Variable capacity tuning is normally used for radio receivers since it is easier to get a better range of coveragel Variable inductance tuning, however, is often used on radio control receiver circuits where tuning is required only to one particular signa. frequency.

We can now reduce the tuned circuit to practical component values. In the medium wave broadcast band the range of frequencies to be covered is from 500 to 1,500 kilocycles/second. (Note here that frequency is far more useful than wavelength for design purposes, which is why the term is used rather than wavelength.)

Typical values of tuning capacitors are from 50 to 500 picafarads and so coil inductance must be selected to match and give a resonant frequency range from 500 to 1,500 kilocycles/second. The inductance of a coil depends on a whole variety of factors—wire size, coil diameter and number of turns, and also whether the coil has an air core or a soft iron core. There are thus a large number of variations possible.

In the case of an air-core coil—a coil wound on a plain circular former of insulating material—100 turns of 38 s.w.g. enamelled or enamelled and double silk covered wire will produce a matching inductance, tapped at 32 turns as shown in Fig. 12.3. This will cover the medium range tuning band in conjunction with a 50–500 pf variable capacitor. By extending the winding to 225 turns this will cover the long wave band as well. Such coils, however, have a relatively

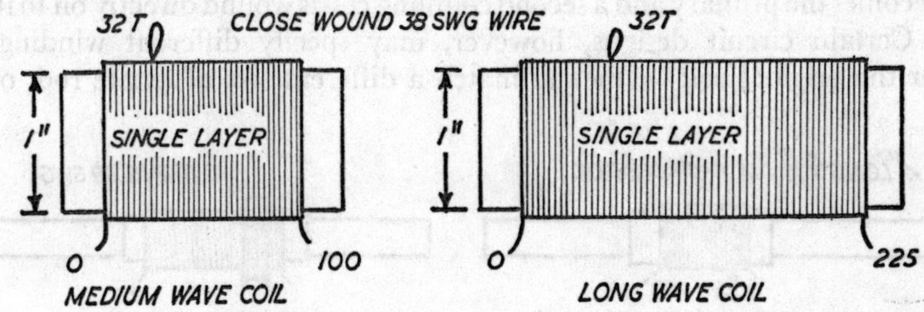

Fig. 12.3. *Medium and long wave coil windings.*

low efficiency or "Q" factor, and this is specially the case with the long wave air-core coil. It would need a long external aerial to give satisfactory reception. The medium wave coil would also need an external aerial, but this need not be so long.

A much better efficiency, or higher "Q" is obtained by winding the coil on a ferrite rod. In this case the coil winding specification depends on the size and type of rod and may also be specific to a particular circuit in order to accommodate a particular form of coupling the tuned circuit to the next stage. Two typical ferrite rod coil windings are shown in Fig. 12.4. The first is a plain tuning coil wound in 38 s.w.g. enamelled wire and the equivalent of the medium wave air-core coil. Because of the improvement in inductance with the iron core (ferrite rod) it will be noted that the coil is both smaller in diameter and length. It is also more efficient than the air-core coil and will work without an external aerial in areas of good reception. The ferrite rod, in fact, takes the place of a separate aerial wire as well as improving the "Q" of the coil. The second diagram shows the same coil with an additional winding added for transformer coupling. This takes the place of the tapping point specified on the single winding coils—the tapping in these cases being the direct coupling point.

The tuned circuit, comprising a 50–500 pf variable condenser and a simple coil wound to the specifications given above will together form a tuned circuit which can be used for *any* radio circuit following and will give medium wave coverage. Only the coupling is likely to differ— i.e. transformer coupling may be preferred to direct coupling to the tapping point on the coil. With transformer coupling the original coil becomes the primary and a second coupling coil is wound directly on to it.

Certain circuit designs, however, may specify different windings for the original coil—either to match a different size of ferrite rod, or

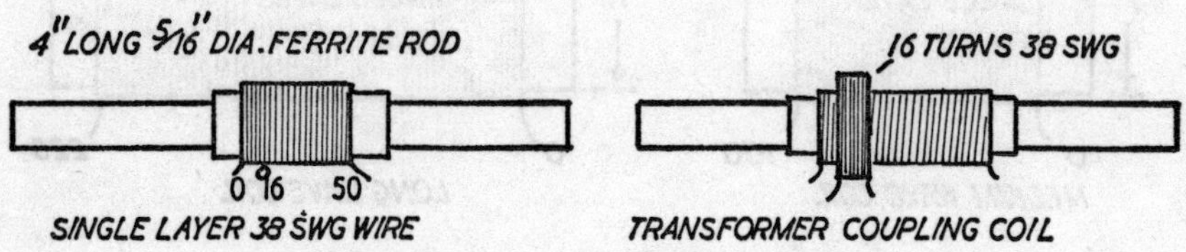

Fig. 12.4. *Two typical ferrite rod coil windings.*

to match a tuning capacitor with a different range. In the case of transformer coupling, a further specification will be given for the coupling coil. Such aerial (or tuned circuit) coils are wound in just the same way—or the standard coils described above could also be used instead with a 50–500 pf capacitor. For coverage of the long wave band the making of coils is a little more difficult since ordinary wound coils do not have sufficient "Q". For best results the long wave coil needs to be wave wound by machine and so such coils are best purchased already wound.

Proceeding to the detector stage this need involve nothing more complicated than a crystal diode to act as a rectifier for the modulated RF signal fed from the tuned circuit. Fig. 12.5 shows a diode detector stage added to a tuned circuit with direct coupling. The combination forms what is a complete crystal receiver which requires no battery. Power is supplied purely by the incoming RF signal picked up by the aerial and selected and amplified in the tuned circuit. With a good external aerial, and in areas of good reception, the signal strength passed by the diode detector can be of sufficient level to produce audible reception in a crystal earpiece or headphones. The capacitor following the diode is simply a by-pass path or filter for RF passing the diode. Its value is chosen so that it offers a *conductive* path for RF but a *resistive* path for AF. Thus all the lower frequency (AF) passed by the diode is directed through the phones. A typical capacitor value for this purpose

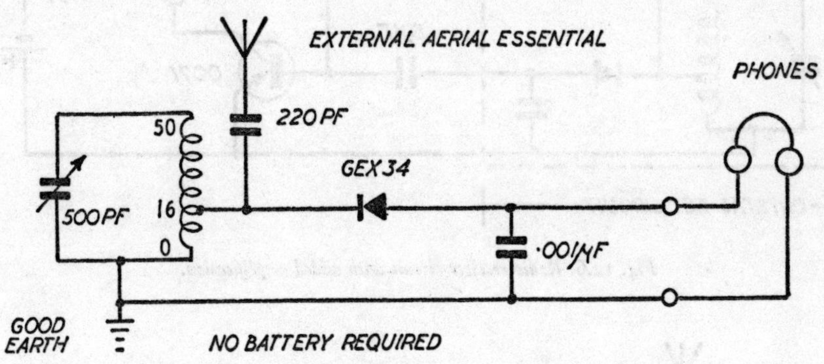

Fig. 12.5. Diode detector stage added to tuned circuit with direct coupling.

is ·001 mf. The return circuit is a wire connection back to the tuned circuit. The performance of the whole receiver will, however, be materially improved if this is also physically connected to a good earth, such as a water pipe. In fact, simple crystal receivers of this type do not work very well without an external earth connection.

Exactly the same circuit up to this stage can be used for better performance by adding amplification, as shown in Fig. 12.6. The transistor used for the amplifier is an OC71 and component values specified have been chosen accordingly. The circuit now requires a 9-volt battery to work the transistor and the output should be good enough to produce a reasonable volume in a high impedance crystal earpiece or phones.

Although very simple and using only a minimum of components, this circuit will not be entirely satisfactory since the transistor is not stabilized. Considerable improvement in this respect is realized by modifying the circuit as shown in Fig. 12.7, the addition of two further resistors and a capacitor making all the difference to stability.

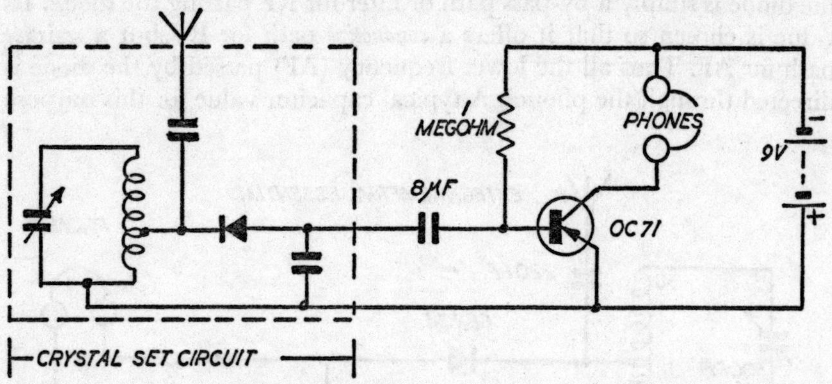

Fig. 12.6. Radio receiver circuit with added amplification.

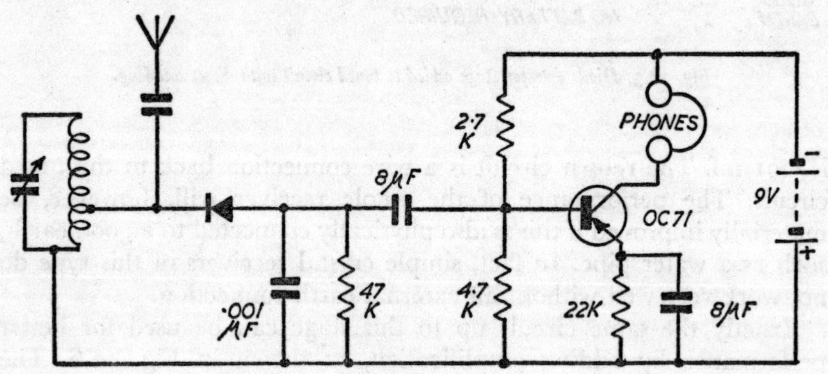

Fig. 12.7. Stabilized radio receiver circuit.

Yet another improvement which can be made is to add a second stage of amplification (Fig. 12.8). This merely replaces the phones in the original circuit with a 4,700 ohm resistor and couples the collector of the first transistor to a second transistor amplifier stage through an

8 mf capacitor. This should make the basic receiver quite workable in areas where reception is poor, and considerably improve the coverage in terms of number of individual stations which can be tuned in. The output, however, will still only be sufficient to operate phones or a crystal earpiece and not a loudspeaker. It could be boosted to "loudspeaker level" by a further stage or two of amplification, but this would not necessarily produce acceptable results. It is likely to show up the limitations of the simple circuit in the matter of sensitivity and selectivity of signal. It is better, therefore, to go to other types of

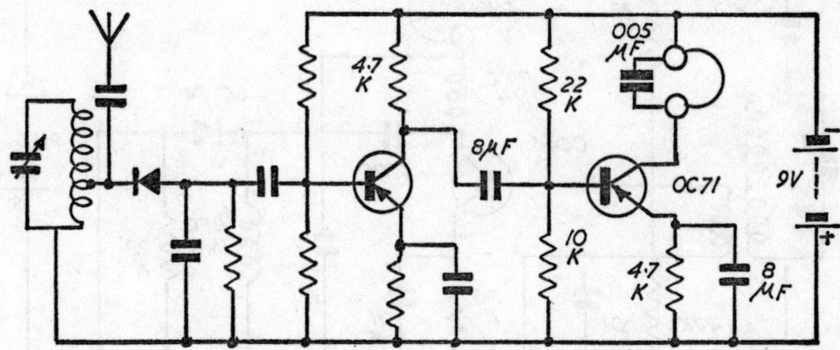

Fig. 12.8. Radio receiver circuit with second-stage amplification.

circuits when loudspeaker working is required, and also where better selectivity and sensitivity is required with earphone working. The main attraction of the basic circuits described are that they are so easy to make and get working; they use only a minimum of components, and also demonstrate very clearly the construction and working of the typical receiver stages.

For improving the performance of a simple transistor receiver without greatly complicating the circuitry the so-called reflex system is often used. This means, basically, that the audio frequency output from the diode is fed back or reflexed into the input. The signal extracted from the tuned circuit is also amplified as an RF signal before being fed to the detector stage in order to overcome one of the basic limitations of the straight tuned circuit-detector circuit. The feedback is merely a matter of circuit connection. Amplification of the tuned circuit output can be achieved quite simply by transformer coupling into a transistor amplifier in front of the detector. The complete basic circuit is then as shown in Fig. 12.9. If required, one or more stages of amplification can be added *after* the detector, just as with the previous circuit. There is, however, a practical limit as to how far this can be carried and show beneficial results. Thus one stage of amplification following the

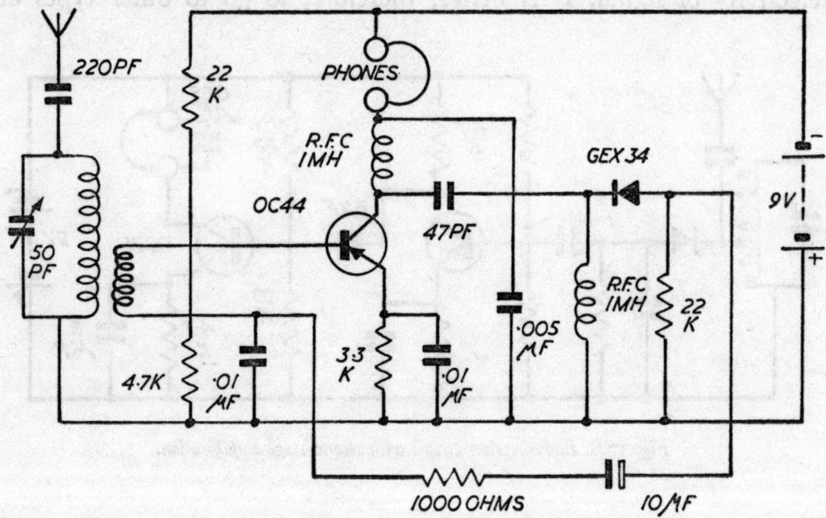

Fig. 12.9. *Simple transistor radio receiver incorporating reflex system.*

detector should be adequate for headphone reception, but two stages might give better results in areas of poor reception. Three stages of amplification following the detector will amost certainly produce poor, noisy and very mixed reception and will still not give good results with a loudspeaker because of this, even though the circuit is now using four transistors.

A circuit design by Mullard for a three transistor receiver capable of working an 80 ohm loudspeaker is shown in Fig. 12.10. This again is a reflexed circuit with part of the output from the diode detector fed back to the first stage. The first transistor operates in a circuit which acts as a regenerative RF amplifier and an audio preamplifier simultaneously. The amplified output from this stage is developed across the choke L2 with part of the signal returned to the aerial circuit through C6 and L1C.

The RF signal output is applied to the diode, which is slightly forward biased to improve its efficiency of working, and the resulting AF signal from the diode is fed back to the base of the first transistor via C7 and L1B. The amplified audio signal is finally fed to the base of the second transistor (driver) via R3 and C5, the variable resistance R3 acting as a volume control. The second transistor works as a low current audio frequency amplifier to drive a Class A output stage. A high impedance loudspeaker forms the load in the collector circuit of the output stage and thus eliminates the need for an output transformer.

The coil windings required are specific to this particular circuit. A Mullard FX2367 ferrite aerial slab is used, with windings as follows:

L1—60 turns of 12 strand 46 s.w.g. wire
L1B—3 turns of 3 strand 46 s.w.g. wire
L1C—4 turns of 3 strand 46 s.w.g. wire

L1B should be interwound with the "earth" end of L1A; and L1C should be interwound with L1A $\frac{1}{4}$" in from the earth end. The choke coil L2 consists of 100 turns of 3 strand 46 s.w.g. wire wound on a tuning slug from a 470 kc/s IF transformer.

A superior performance can be achieved with a superhet circuit and most commercial receivers are of this type. The superhet is, however, a considerably more complex arrangement involving many more components and the necessity of aligning the complete circuit for satisfactory working. Its description and use is, therefore, outside the scope of this book, although circuits of this type suitable for home construction are available in kit form.

Although still a basically simple receiver circuit this particular design is readily capable of working an 80 ohm 2" or 3" diameter loudspeaker to give a good output volume in areas of reasonable reception. The circuit is also reasonably simple to build and lends itself to printed circuit assembly. Alternatively, it can be built by any of the methods described in Chapter 2. Care should be taken to ensure that the electrolytic capacitors are connected the right way round.

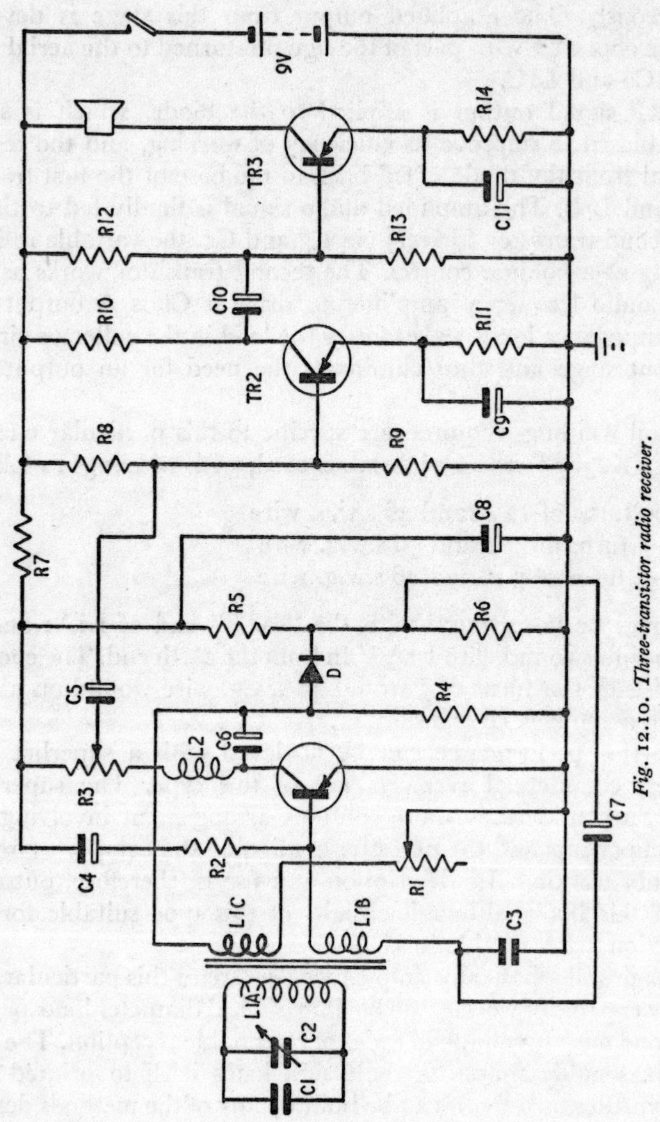

Fig. 12.10. Three-transistor radio receiver.

Components list:

Resistors

R1	50kΩ	linear pot.
R2	560kΩ	
R3	5kΩ	log. pot.
R4	4·7kΩ	
R5	150kΩ	
R6	10kΩ	
R7	2·2kΩ	
R8	33kΩ	
R9	4·7kΩ	
R10	1·5kΩ	
R11	470Ω	
R12	680Ω	
R13	100Ω	
R14	10Ω	

Capacitors

C1	12pF	
C2	365pF	Jackson type 01
C3	0·02μF	Mullard C296AA/A22K
C4	100pF	
C5	10μF/16V	Mullard C426AM/E10
C6	330pF	
C7	10μF/16V	Mullard C426AM/E10
C8	40μF/16V	Mullard C426AM/E40
C9	100μF/4V	Mullard C426AM/B100
C10	40μF/16V	Mullard C426AM/E40
C11	320μF/2·5V	Mullard C426AM/A320

Inductors

L1A 60 turns of 12 × 46 s.w.g. Litz
L1B 3 turns of 3 × 46 s.w.g. Litz
L1C 4 turns of 3 × 46 s.w.g. Litz

L1B should be interwound with the earthy end of L1A, and L1C should be interwound with L1A, ¾″ from the earthy end.

The Ferrite aerial slab should be Mullard FX2367.

L2 consists of 100 turns of 3 × 46 s.w.g. Litz wound on a tuning slug from a 470kc/s i.f. transformer.

Transistors and Diode

TR1 Mullard AF117
TR2 Mullard OC71
TR3 Mullard OC81
D1 Mullard OA70

CHAPTER 13

RADIO CONTROL

A RADIO control system comprises, essentially, a transmitter, a receiver and an actuator. Transmitter and receiver together provide a radio link whereby movement of a switch on the transmitter is translated in the form of a switching action by the receiver some distance away. The actuator is some form of device for transforming the switching action of the receiver into "muscle power", as it were, to provide some type of mechanical movement. Thus the complete system may be employed for remote operation of the controls on a working model, such as an aircraft or boat; for functional duties like opening a garage door or operating a camera shutter by signalling from a distance; for conjuring tricks and stage effects operated by remote control, and so on.

It will be noticed that all these applications of radio control work on a simple signal basis and do not include the transmission of speech or music. Technically there is nothing to stop a radio control circuit which works on a signal being modified to transmit speech, but this is illegal unless the operator possess a special Radio Amateur's licence and operates on a specified frequency. For the same reason, although there are numerous walkie-talkie radio sets sold, both as surplus and as new manufactured items imported (notably from Japan), it is illegal to operate such sets in this country.

The field of radio transmission is thus restricted for simple control purposes only to anyone who has not qualified for a "ham" licence—the only exception being that an educational licence can be obtained to operate a low powered transmitter-receiver link for demonstration purposes in schools and colleges where the range can be restricted. For radio *control* systems (which prohibits the transmission of speech or music), a special frequency band is allocated and all such transmitters must work within this frequency. It is also necessary to obtain a licence before such sets can be operated, but this is available on application, without having to pass any examination, and costs £1 for a period of five years. The allotted frequency for such radio control work is a comparatively narrow band of 26·96 to 27·28 megacycles/second, equivalent to a wavelength of something like 10 metres (32 feet).

The subject of radio control is quite complicated and tends to become highly specialized when applied to models. For the purpose of this book, therefore, we shall deal only with basic single-channel systems; how they work and how such transmitters and receivers can be made. Those readers particularly interested in the subject are recommended to seek further information in the many books published on radio control for models and radio control systems.

In its simplest possible form a radio control transmitter need only provide an RF signal at the required (legal) frequency of 27 megacycles/second. A matching receiver can then be tuned to this signal and the radio link between the two established by switching the transmitter signal on and off. This will result in a corresponding change in the receiver (in practice a current change) which can be used to operate a switching circuit controlling the actuator. The actuator is a separate electro-mechanical device with a mechanical output, this output providing the control movement required.

As far as the transmitter is concerned, all that is really needed is an oscillator tuned to a frequency of 27 megacycles/second and with the output fed to an aerial. Power output is limited to a maximum of 1·5 watts under G.P.O. regulations, although the actual output used on practical transmitters is usually very much less, especially in the case of hand-held transmitters which have to work off dry batteries. The actual efficiency of the transmitter circuit, which determines the power output, may vary widely with different circuits and can be as low as 10 per cent. Overall efficiency, too, is very much affected by the type and efficiency of the aerial used. Nevertheless, with quite low power outputs, any good transmitter should be capable of a ground-to-ground range of several hundred yards, matched to a suitable receiver, with a corresponding ground-to-air range of several times this figure. This is more than the range which can be usefully employed in practice since a model aircraft, for example, is so small at a distance of half a mile or so away as to be impossible to control effectively at that range in any case. With model boats the practical range required is very much shorter. One can, therefore, design down to a range, if desired to considerably simplify the circuitry involved. It is more usual to find this done with receivers rather than transmitters, however.

A triode valve coupled to a tuned circuit will readily perform as an oscillator, when a complete transmitter need consist only of a minimum number of components, as shown in Fig. 13.1. The only critical factors involved are the values of the inductance and capacitance in the tuned circuit to establish oscillation at 27 megacycles. It is usual with circuits of this type to employ an air-core coil for the inductance with a variable capacitance to adjust the resonant frequency to the final figure required.

A suitable air-core coil can be made from 18 s.w.g. or 16 s.w.g. enamelled wire, as shown in Fig. 13.2. Eight full turns are wound on to a suitable

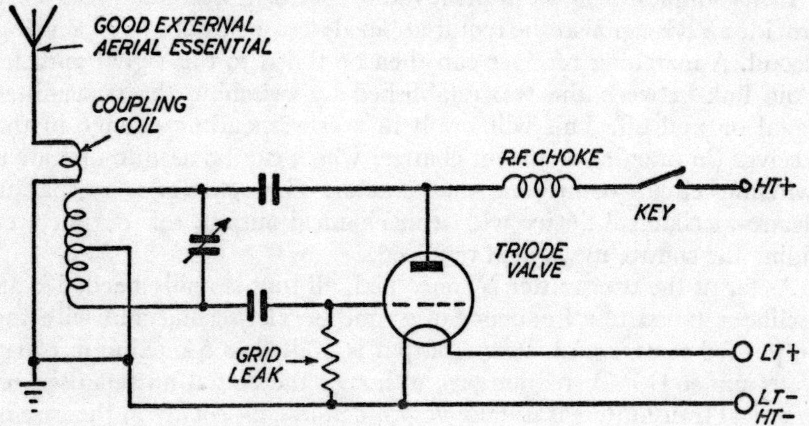

Fig. 13.1. *Simple radio transmitter.*

former $\frac{7}{8}''$ diameter (a piece of dowel for example) and the coil then removed from the former and pulled out to a length of $1\frac{1}{2}''$. It is then securely fastened to tags or through holes drilled in the panel carrying the transmitter circuit. A matching size of capacitor for tuning to the 27 megacycle frequency is 3–30 pf. This then completes the tuned circuit, except for the coupling coil to the aerial. This is a single loop of 18 or 16 s.w.g. enamelled wire, covered with sleeving, and wound around the centre of the main tuning coil. Assembly of the rest of the circuit then follows quite logically, mounting the valve in a suitable valve base.

There are several limitations with this simple circuit, notably a tendency towards instability or change of frequency for a variety of reasons, but one of the main snags is setting up the correct frequency of oscillation initally. The circuit, and thus the signal frequency, is tunable over quite a wide range via the variable capacitor. To operate within the legal frequency band it needs very carefully setting up with reference to some instrument which is accurately calibrated as regards frequency.

The simplest way to do this is to tune the transmitter to a superhet receiver, if available. A superhet receiver will be set up to, and only work on, one spot frequency in the 27 megacycle band. Tuning the transmitter to work such a receiver will, therefore, set it up to that particular spot frequency. It is no good trying a similar method with a

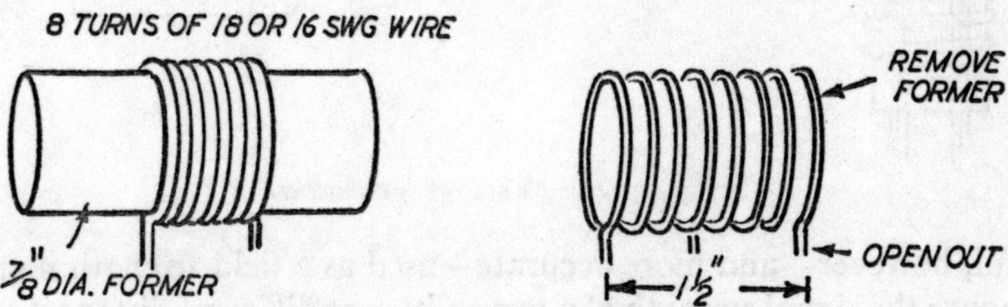

Fig. 13.2. *Making an air-core coil.*

super-regen receiver, which has already been tuned to another 27 megacycle transmitter, since the response of such receivers is too broad. Thus the final tuning point established on the transmitter may well be outside the 27 megacycle band although the receiver is apparently tuned *to* the 27 megacycle band. Lacking a superhet receiver to work to, the only satisfactory way of setting up the transmitter tuning is against an accurately calibrated absorption wavemeter.

A simple form of absorption wavemeter capable of detecting and indicating signals in the 27 megacycle region is shown in Fig. 13.3 and consists of little more than a tuned circuit of a coil and capacitor with a smaller (one turn) coupling coil for output. This coupling coil can be connected to a 1·5 volt flashlamp bulb, or to a 0–1 milliammeter or 0–100 microammeter. When tuned to the signal by adjustment of the capacitor the bulb will show the strongest light, or the meter the greatest reading. If initially tuned to a crystal stabilized transmitter, therefore, the adjustment can be left alone and another transmitter tuned *to* the wavemeter.

A somewhat more efficient and useful field strength meter is shown in Fig. 13.4 employing a simple tuned circuit again, followed by a diode

detector plus transistor amplification to give readable deflections on a
0–5 milliammeter. Again if set up initially against a stabilized transmitter signal it can be used to tune further transmitters. It will be most

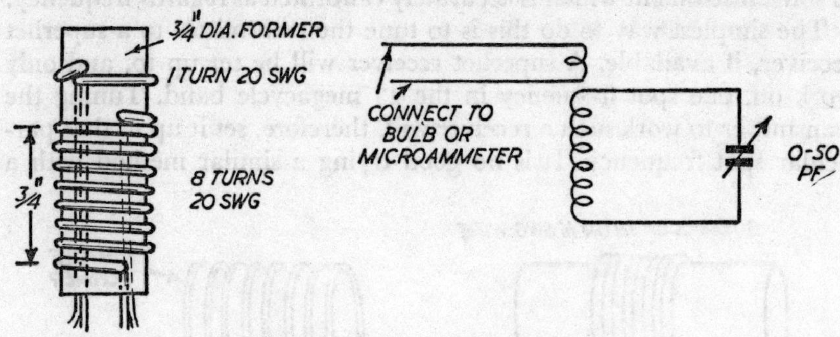

Fig. 13.3. *Simple absorption wavemeter.*

useful, however—and more accurate—used as a field strength meter to
compare the signal strength of a transmitter at different distances.

Instead of a milliammeter, phones can be used in the circuit at this
point to actually listen to the transmitter signal receiver for an audible
rather than visual method of tuning. This is rather less accurate than
working to a meter reading since it is difficult to distinguish maximum

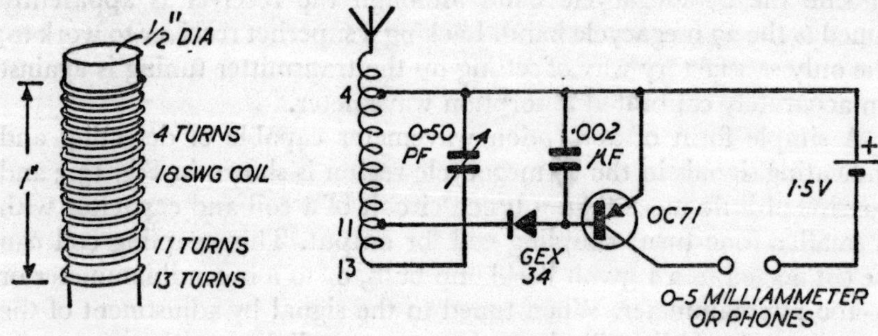

Fig. 13.4. *Field strength meter.*

volume. The phones (or crystal earpiece) should be of the high impedance type.

Besides being very simple to make, a field strength meter of this type
is a valuable piece of equipment for any serious radio control enthusiast.

Transmitter circuit stability is greatly improved if a *crystal* is incorporated, the transmitter frequency then becoming, in effect, crystal controlled. This implies the use of a special quartz crystal with a resonant frequency at a suitable spot in the 27 megacycle band, or a sub-harmonic of the required frequency. This can be connected directly into the basic circuit, as shown in Fig. 13.5.

A rather more reliable transmitter circuit—and one, incidentally, which is almost impossible not to work correctly if properly wired up—

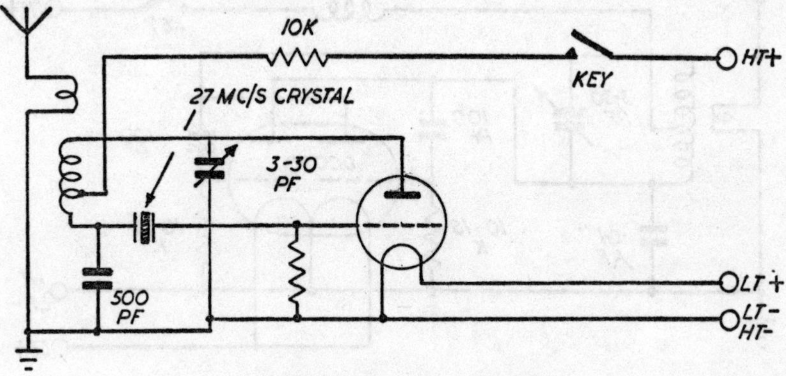

Fig. 13.5 *Incorporation of crystal into transmitter circuit.*

is a push-pull Hartley oscillator, commonly known as a cross-coupled circuit. Basically this employs two triode valves, cross-coupled anode to grid, but a suitable double triode valve can be obtained in a single valve envelope (for example a DCC 90 or 3A5). The complete circuit is then shown in Fig. 13.6, using the same tuning coil as previously described. There are no difficulties at all with regard to construction and the whole can be assembled on a Paxolin panel measuring about 4″ × 3½″. This, in turn, can be mounted in a suitable case large enough to accommodate the 90 volt HT battery and 1·5 volt LT battery. These can be large size batteries for maximum life if the transmitter is to be ground-standing; or smaller batteries (90v B.136 for example) to save weight if the transmitter is to be hand held. With a ground-standing transmitter an 8 ft. aerial is recommended.

The valve transmitters described are simply oscillators giving a continuous 27 megacycle or RF signal when switched on and nothing when switched off. Their performance can be further improved by adding an amplifier circuit (particularly in the case of the single valve oscillator) to produce what is known as a MOPA (master oscillator power amplifier) circuit. This is usual with modern commercial radio control transmitters. It is also more usual in modern transmitter design to utilize a modulated signal rather than a basic RF signal which is simply switched on and off. This is produced by superimposing a tone or low frequency

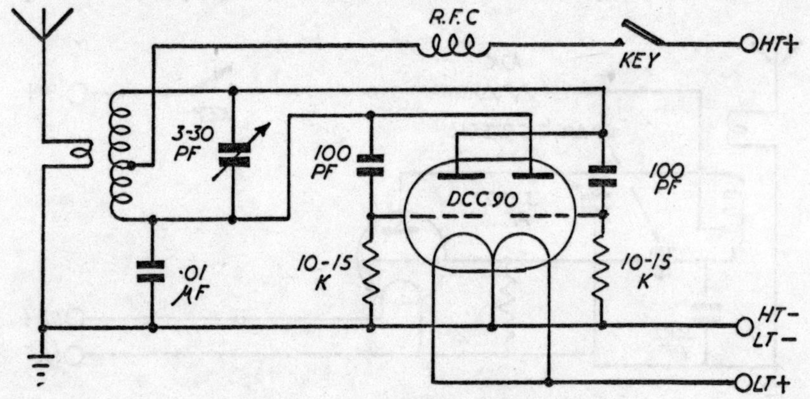

Fig. 13.6. *Cross-coupled transmitter circuit.*

(AF) signal on the basic RF or carrier wave. It is known as a tone transmitter as distinct from a carrier transmitter. In the case of a tone transmission the carrier can be left switched on all the time and the tone switched on to modulate the carrier for signalling purposes (although in some circuits carrier and tone are keyed simultaneously to save the battery from current drain in supplying the carrier circuit when no signal is required).

A tone transmitter is simply a carrier transmitter fitted with an additional tone generator or modulator. There are many ways of making a tone generator, but a multivibrator circuit is most widely favoured using either a pair of triode valves or pair of transistors. In the former case a double triode can be used so that both valves are accommodated in one envelope. Both types of tone generators are shown in Fig. 13.7 and can be applied to virtually any simple carrier transmitter circuit, although the method of coupling may be specific to the circuit used. The transistor tone generator is the simplest to make and in the case of the basic transmitter shown in Fig. 13.1 would simply connect to the grid of the valve via an RFC choke.

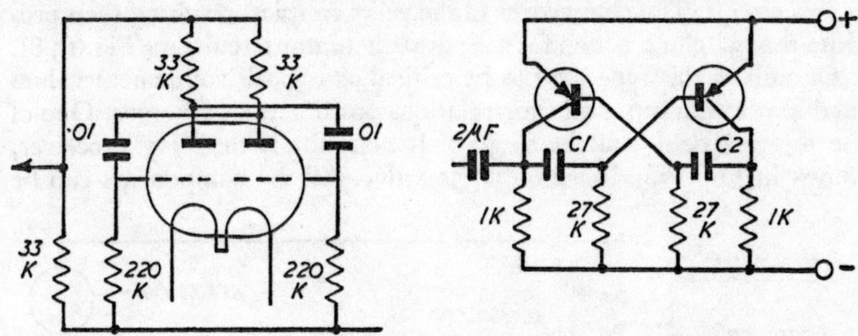

Fig. 13.7. *Two tone generator circuits, using a double triode valve (left) or two transistors in multi-vibrator type circuit (right).*

Suitable values for C_1 and C_2 (Fig. 13.7) are $\cdot 1\mu F$. If a higher frequency tone is required, decrease this value, and vice versa for a lower tone.

The function required from the receiver circuit is to pick up the signal from the transmitter and convert this into a useable form of electrical energy—usually a marked change in current flowing through the receiver circuit. This it has to do with a minimum of components so that the complete receiver can be reduced to the smallest possible physical size and weight. In practice this virtually restricts the choice of circuit to super-regenerative or superhet, both being essentially similar but the latter employing an additional "front end" to the circuit. The superhet is not a suitable type for amateur construction (except from kits employing printed circuit assembly), which leave only the super-regenerative or super-regen circuit to be considered.

A super-regen receiver circuit is, in fact, a simple oscillator, but differing from a conventional oscillator in that the oscillations are self-quenched. That is to say the receiver oscillates in bursts only, the frequency of the bursts being known as the quench frequency. In practice the circuit is adjusted until it is just on the point of oscillating. In this state it is very sensitive to small voltages which may be applied to the grid of the valve, such as an incoming signal from a transmitter. The result is that the appearance of such a signal can cause a change of condition from non-oscillating to self-quenched or intermittent oscillation, with a resulting change in anode current of 2 to 3 milliamps. This current change is quite sufficient to operate a sensitive relay connected in this circuit. The changeover of the relay contacts resulting then provides the switching action for a separate actuator circuit (see Fig. 13.8).

Circuits of this type tend to be critical as regards component values used and setting up, and the variations possible are numerous. One of the most successful of the single valve circuits is the "Ivy" receiver, shown in Fig. 13.9, based on a 354 valve. All the components can be

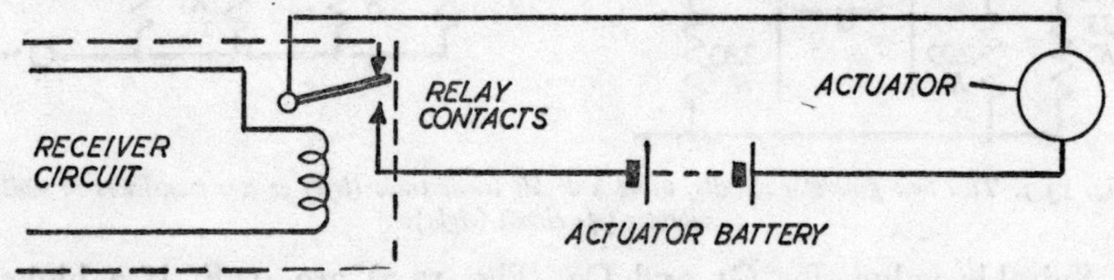

Fig. 13.8. Super-regen receiver circuit connected to relay circuit.

accommodated on a Paxolin panel about $3\frac{1}{2}'' \times 2''$ with the valve plugging into a B7G valve base. There is also sufficient space on this size of panel to accommodate a conventional 4–5,000 ohm sensitive relay. The circuit is designed for carrier wave operation and thus, in conjunction with the carrier wave transmitter described earlier, provides the simplest (and cheapest) combination of transmitter and receiver for model radio control work.

The only critical components involved in this circuit are the tuning coil, sensitivity coil and quench coil. Unlike the domestic radio receiver (see Chapter 12), the tuned circuit of radio control receivers is normally tuned by a variable inductance associated with a fixed capacitor value. This means winding the coil, on a coil former which is subsequently fitted with an iron dust core—final tuning or adjustment of inductance being made by screwing this core in or out of the former.

The tuning coil for this circuit is made by winding 9 close turns of 28 s.w.g. enamelled wire on to a $\frac{1}{4}''$ diameter polystyrene coil former, taking out a loop for a tapping point and then winding on another 9 close turns to complete the coil.

The sensitivity coil is also wound on a $\frac{1}{4}''$ diameter polystyrene coil former with an iron-dust core and this time consists of 10 turns of 28 s.w.g. enamelled wire. These turns should be located at the bottom of the former. Two ¾″ squares of thin ply are then cut, drilled to fit over the remaining plain part of this former, and cemented in place to form a bobbin. On to this bobbin space are wound 600 to 650 turns of 40 s.w.g. double silk covered wire to complete one winding of the quench coil;

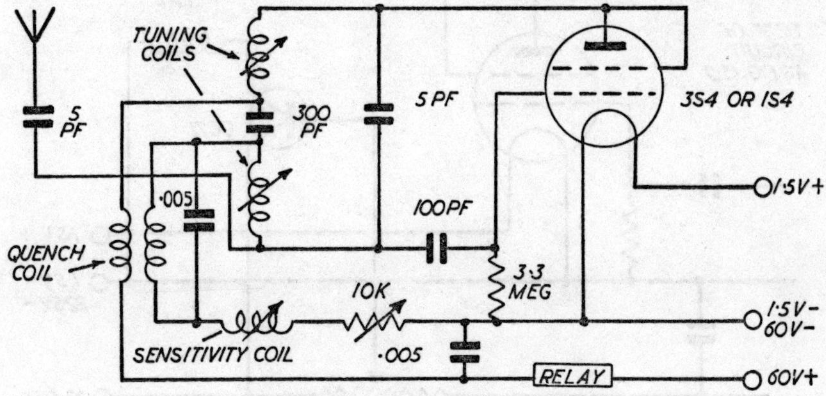

Fig. 13.9. 'Ivy' receiver circuit.

and on top of this a similar number of turns of the same wire for the second winding of the quench coil. Alternatively the two windings for the quench coil can be made on a separate bobbin or former and the coil separately mounted.

Having made these coils it is then a matter of fixing them to the Paxolin panel together with the valve base and relay and wiring in the remaining components to complete the circuit. A high tension supply of 45 or 60 volts is required (the higher voltage will give greater current change) and also a low tension battery supply of 1·5 volts.

The receiver is best set up by inserting a 0–5 milliammeter in the HT battery lead. With the batteries connected, the slug in the tuning coil former is screwed one way or the other until the current rises to a maximum, which should be about 3 to 3·5 milliamps with 60 volts HT. Then the core of the sensitivity coil is adjusted until the current drops to about 2 milliamps.

Switch on the transmitter and return to the tuning coil slug and adjust for *minimum* current. Switching off the transmitter should then restore the current to a maximum value. Adjust tuning and sensitivity core slugs, as necessary, to achieve a positive current fall each time the transmitter is keyed. It is assumed that the relay is already set up to a suitable operating current, if not it will have to be adjusted to pull in at some current a little below the maximum current of the receiver and

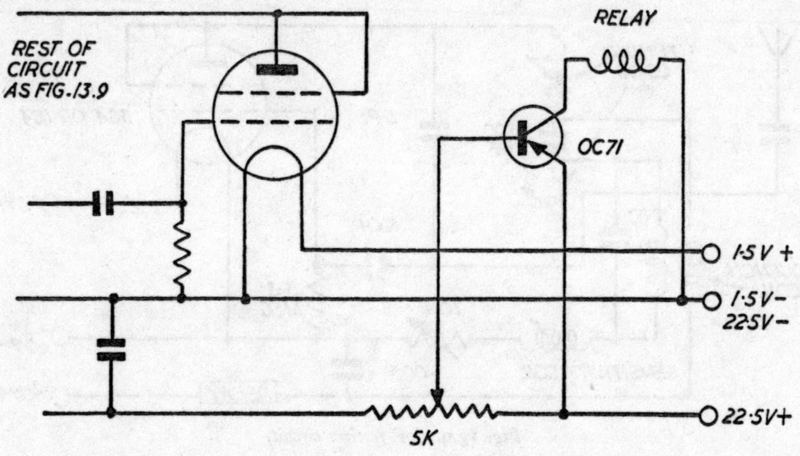

Fig. 13.10. *Circuit as in Fig.* 13.9 *adapted to transistor amplification.*

drop out at about ·2 milliamps below the pull in current. The current fall on response to the transmitter signal will then pull the relay in strongly; and switching off the transmitter signal will cause the relay to drop out. The sensitivity control is "backed off" as necessary to achieve stable operation of the circuit.

Modern radio control receivers are almost invariably of transistor type, either all-transistor or with a valve detector or transistorized throughout. The immediate advantage is that apart from reducing the physical size of the receiver, transistor circuits require much lower battery voltages.

The circuit just described can be adapted to transistor amplification with a considerable improvement in performance and a reduction in HT battery requirements to 22·5 volts. This modified circuit is shown in Fig. 13.10, incorporating an OC71 transistor. All the coils and other component values are as before, with the exception of the added 5,000 ohm variable resistor. This is used to regulate the current flow through the relay, that is to set up the top or standing current. A similar potentiometer can also be incorporated in the original circuit with advantage for the same purpose.

Yet another modification which can be made to the original circuit is to incorporate two further stages of transistor amplification when the current change in the final stage can be as high as several hundred milliamps, depending on the load. This is a sufficient current change to operate an actuator direct, dispensing with the relay as a switching

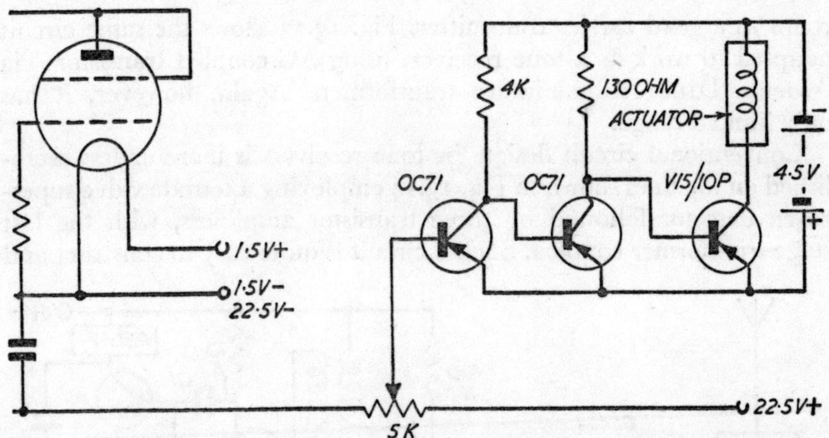

Fig. 13.11. *Additions to circuit in Fig.* 13.9 *needed to make a relayless receiver.*

device for a separate actuator circuit. It is known as a relayless receiver, and the necessary circuit additions are shown in Fig. 13.11.

All-transistor receivers tend to be a bit tricky for amateur construction, largely because transistors are not like valves and behave somewhat differently in similar circuits. However they can readily be used with a crystal diode detector in a practical receiver circuit, although such a detector circuit will not have the same sensitivity, and thus the range, as a valve super-regen detector.

Fig. 13.12 shows a very simple circuit of this type, which is more of a novelty than a practical receiver since it can only be expected to have a range of a few yards. It is, however, a very simple receiver which could be used for operating a remote control device within the same

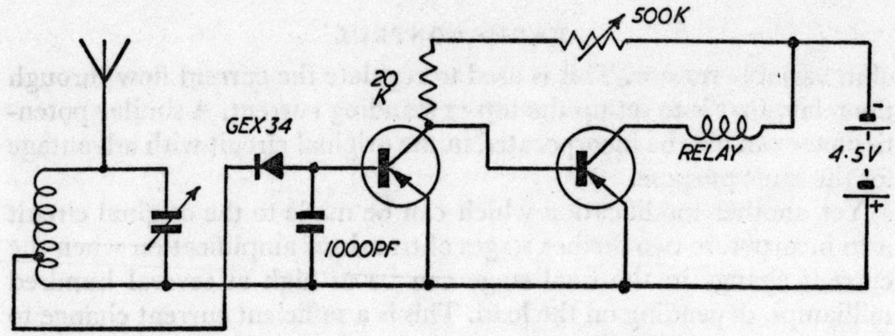

Fig. 13.12. Simple detector circuit incorporating crystal diode.

room as a good carrier transmitter. Fig. 13.13 shows the same circuit adapted to work as a tone receiver, using AC coupled transistors via Ardente D1001 subminiature transformers. Again, however, it has only limited range.

Conventional circuit design for tone receivers is more or less established on the lines shown in Fig. 13.14 employing a tetrode valve superregen detector followed by three transistor amplifiers, with the last stage transformer coupled. Such a circuit is quite easy to construct and

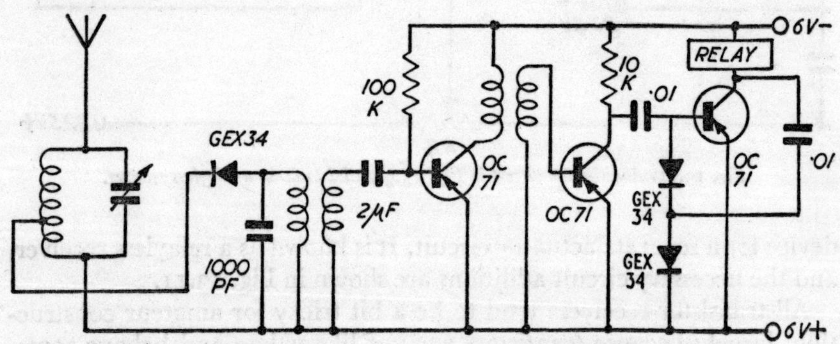

Fig. 13.13. Adaptation of circuit in Fig. 13.12. to work with tone receiver.

all component values are given. It is a superior type of receiver to the ones previously described and also avoids the use of a quench coil. Only one adjustment is required for tuning—the core of the tuning coil. It also has the characteristic of operating the other way round; that is receipt of modulated tone transmitter signal is accompanied by a current rise. The normal standing or idling current is high when switched on, falling to a low value when the transmitter *carrier* is switched on. Current then rises again in the receiver circuit every time the transmitter *tone* is signalled.

It will also be appreciated from this that there is no real advantage in keying carrier and tone simultaneously on a tone transmitter to save the transmitter batteries from drawing current all the time. It is far more practical to save the *receiver* batteries by leaving the transmitter

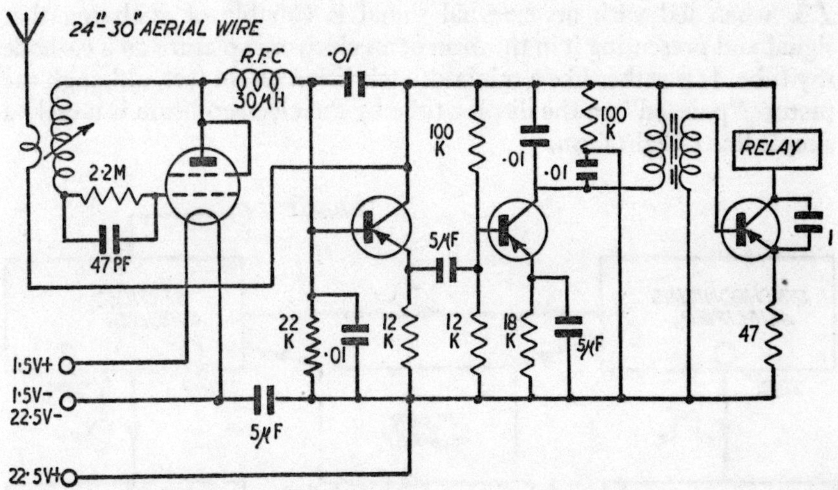

Fig. 13.14. *Conventional circuit design for tone receiver.*

carrier on all the time and signal by keying the tone circuit only. Also, of course, a tone receiver will only operate properly in conjunction with a tone transmitter.

The all-transistor tone receiver is a little more tricky, but it can be produced as a working circuit. The main advantages are that the size of the receiver can be reduced to an absolute minimum, and also only a single low voltage battery is required. It is also quite easy to make the receiver a relayless type by adding sufficient stages of transistor amplification and thus further save size and weight. The main trouble lies with the limitation of the transistor as a detector, compared with a valve, necessitating the use of a crystal diode in the circuit.

There are numerous variations possible on such all-transistor circuitry, but only the minority are really successful.

CHAPTER 14

THE OSCILLOSCOPE

A N OSCILLOSCOPE is a relatively complex electronic device which when fed with an external signal is capable of analysing that signal and presenting it in the form of an electronic picture on a cathode ray tube. It is rather like a miniature television set, in fact, although the picture "painted" on the display tube by the electron beam is merely a simple line of light form.

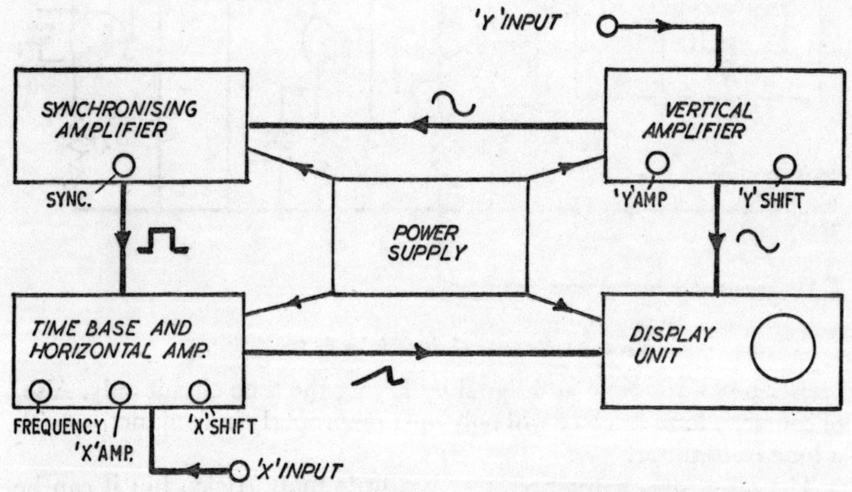

Fig. 14.1. *Layout of simple oscilloscope.*

Basically a simple oscilloscope comprises four separate sources, all inter-related and all drawing their source of power from a common power source (Fig. 14.1). The circuits are (i) the vertical amplifier; (ii) the synchronizing amplifier; (iii) the time base; and (iv) the display unit. All are normal electronic circuits, except that the display unit contains the cathode ray or display tube with its associated focus and brilliance controls.

The vertical amplifier consists essentially of a multi-stage, high-gain amplifier system with two main controls—the "Y" shift and "Y" sensitivity or amplification. The "Y" shift has the capability of deflecting the spot on the cathode ray tube in a vertical or "Y" direction. The "Y" sensitivity merely controls the degree of amplification of signals being fed to the "Y" deflection plates of the cathode ray tube. The circuit is normally designed to accept only alternating voltages as input ("Y" input), but some more complex oscilloscopes make provision for accepting and amplifying direct voltage inputs.

The time base comprises a generator capable of producing a saw tooth wave form, the frequency of which can be varied over a suitable range. When this is fed to the "X" plates of the cathode ray tube this causes the spot to be swept across the screen at a constant speed, the speed being governed by the frequency of the wave form. Thus, essentially, the time base provides a time scale which can be adjusted, as necessary, to measure various electrical waveforms, which are time dependent. The frequency of the time base, and thus the timing range, can extend from a few cycles per second up to many megacycles per second. With a simple oscilloscope, however, the time base range is usually more restricted, for example, from 10 cycles/second up to about 25 kilocycles/second.

Provision is also usually made on oscilloscopes to feed external signals to the "X" plates of the cathode ray tube ("X" input). The time base circuit may also incorporate an amplifier to amplify these signals, but this is by no means universal, or necessary. Thus the controls associated with the time base circuit are an "X" shift control which can deflect the spot horizontally across the screen; a frequency control which varies the frequency of the saw tooth wave form; and an "X" amplitude control where amplification of an "X" input is provided. There is also a simple switch which selects either the time base waveform or the external signal to be applied to the "X" plates of the cathode ray tube.

The synchronization amplifier is merely a conventional amplifier circuit inserted between the time base generator and the vertical amplifier. Its function is to ensure that the time base sweep and the input signal to the "Y" deflection plates can be synchronized—that is applied at the same instant of time. In practice, this means that the trace can be "locked" or held still on the screen.

The display unit itse lfmerely contains the cathode ray tube and its

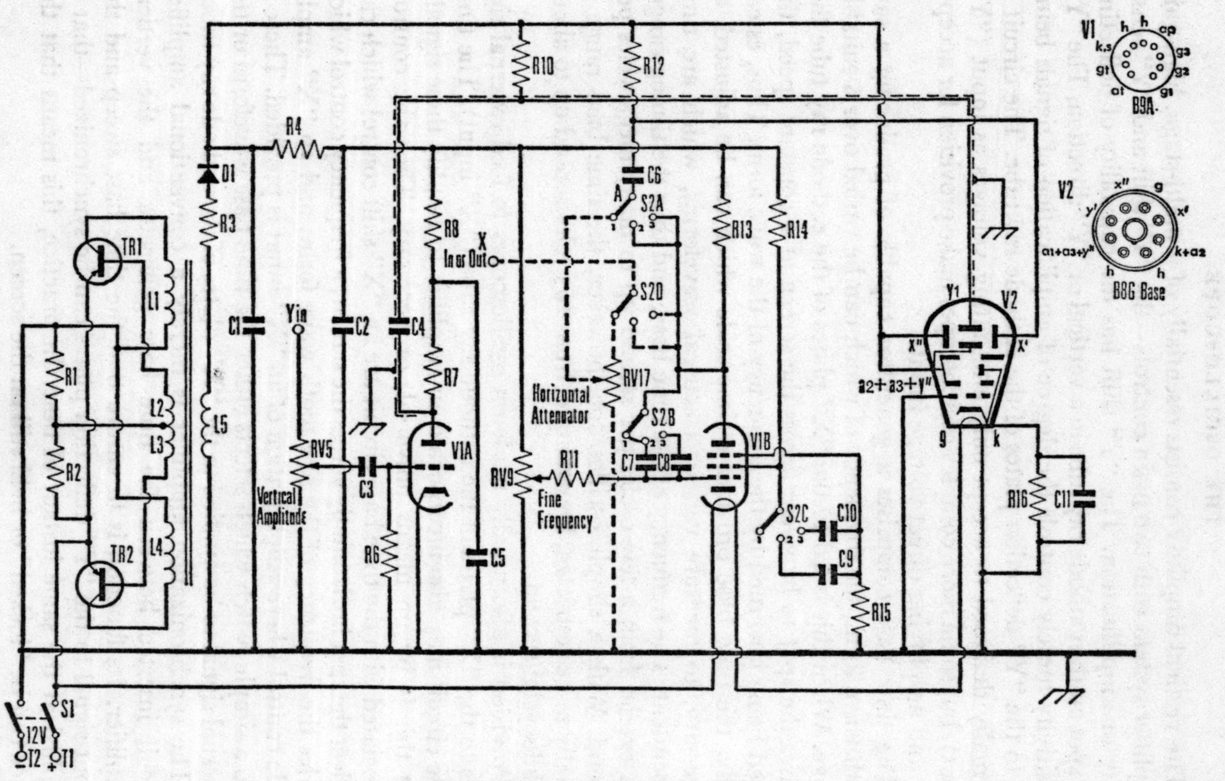

Fig. 14.2. Mullard 12 volt oscilloscope.

associated focus and brilliance controls. These controls are normally adjusted to give minimum spot intensity which is still visible (to avoid "burning" of the screen); and minimum spot size (for the most clearly defined trace).

The power unit is a purely conventional circuit, except that it has to supply high tension for the valves; extra high tension for the anode of the cathode ray tube, and a low tension supply for the cathode ray tube filament and all the valve filaments.

A conventional oscilloscope represents a complicated and expensive circuit to construct, and invariably requires a mains supply for working. The extra high tension may then involve voltages of the order of 500v or greater. However, Mullards have produced a circuit design suitable for amateur construction which operates from a single 12 volt DC supply and a minimum of components. It incorporates no shift controls, no brilliance controls and no focus control and thus has certain limitations as regards quantitative measurement. Nevertheless it is a thoroughly worthwhile project for the electronic hobby enthusiast since it does provide him with a minimum cost oscilloscope which really works.

The complete circuit design is shown in Fig. 14.2. The extra high

Fig. 14.2

Components list:

R1	1·5kΩ		
R2	390Ω ½ watt		
R3	10kΩ		
R4	33kΩ		
RV5	500kΩ linear potentiometer		
R6	8·2MΩ		
R7	220kΩ cracked carbon		
R8	47kΩ cracked carbon		
RV9	2MΩ logarithmic potentiometer		
R10	2·2MΩ	R14	100kΩ
R11	220kΩ	R15	1MΩ
R12	2·2MΩ	R16	1MΩ
R13	47kΩ		
RV17	(if required – see text) 500kΩ linear potentiometer		

All resistors are ¼W, 10% unless otherwise stated.

C1	0·47μF	400V min. wkg.
C2	1μF	400v min. wkg.
C3	0·22μF	150V min. wkg.
C4	0·1μF	500V „ „
C5	0·1μF	150V „ „
C6	0·1μF	400V non-inductive
C7	0·015μF	150V „ „
C8	1000pF	150V „ „
C9	0·015μF	150V „ „
C10	1000pF	150V „ „
C11	0·22μF	miniature
TR1, 2	Mullard OC29 power transistors	
D1	Mullard OA211 or BYX10 rectifier diode	
V1	Mullard ECL80 triode pentode	
V2	Mullard DH3-91 cathode ray tube (300mA heater version)	
S2	Three-way, four-pole rotary switch	
B9A	Case (for valve)	
B8G	Case (for cathode tube)	

tension voltage required for the anode of the cathode ray tube is obtained from transistors TR1 and TR2 in conjunction with a transformer making up a DC converter circuit. The output is half wave rectified by diode D1 with C1 acting as a reservoir capacitor. This yields 400 volts extra high tension for connecting to the cathode ray tube anode. A 300 volt output is also obtained from this circuit through the filter components C2 and R4, used as HT supply for the single triode-pentode valve V1. The triode section of this valve is used as a vertical amplifier, the signal being applied to the grid of this valve via RV5 which acts as a gain control. The pentode section of the same valve is used as the time base generator, the frequency of the saw tooth wave form being governed by capacitors C8, C10 or C7 and C9 in conjunction with the fine frequency control RV9.

The "X" input terminal in the circuit is connected via RV17 which acts as an amplifier. This connection, in conjunction with switch S2, ensures that a saw tooth wave form is available at the terminals when the time base is operating. When the time base is switched off, then this terminal simply accepts an external signal as "X" input. RV17 can be omitted from the circuit if the input signals are likely to be of sufficient magnitude as to not need amplification (attenuation).

Component positioning is not critical, except that the transformer should be located as far away from the cathode ray tube as possible, and perhaps shielded. This is to prevent transformer hum from interfering with or modulating the trace. Otherwise, no difficulty should be experienced in planning a suitable layout with all controls, terminals and switches and the cathode ray tube mounted on a vertical panel forming the face of the instrument; and the remaining components on a suitable base or chassis at right angles to the face. The whole, in fact, can be contained within a case size of 4″ × 4″ × 3″ with the cathode ray tube projecting about 2″ from the face. Because it is liable to accidental damage, the projecting length of tube should be protected with a suitable shield or cover.

Construction of a suitable transformer for the DC converter stage has already been described in Chapter 9. A full list of other components is given in the caption under the circuit diagram (Fig. 14.2).

For the more experienced amateur and the professional wanting to make an instrument for servicing, Mullard have designed a students' constructional oscilloscope. It is a precision instrument of modern design adapted to unit construction and capable of giving an excellent performance at minimum overall cost. Full details of construction are given in a booklet published by Mullard Educational Service, Mullard House, Torrington Place, London W.C.1.

The oscilloscope is, essentially, a highly sensitive voltmeter and thus

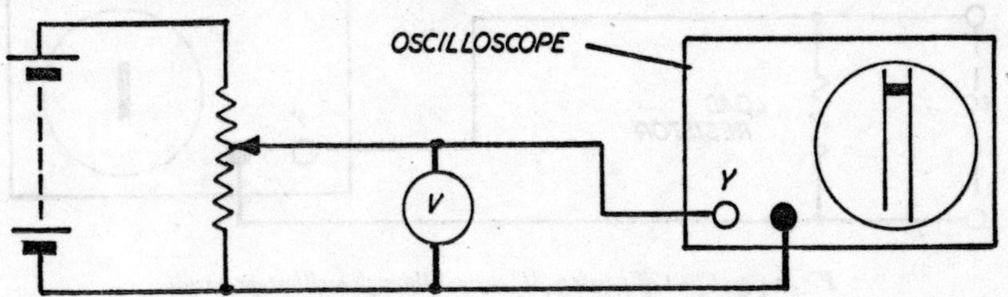

Fig. 14.3. Oscilloscope giving voltmeter reading.

provided the vertical amplifier will accept DC it can be used both to demonstrate and measure DC phenomena. Thus with the circuit shown in Fig. 14.3 connection to the "Y" input variation on the input voltage via adjustment of the 100 K potentiometer will produce a proportionate deflection of the spot in a vertical direction. The oscilloscope "Y" gain control, where fitted, can be used to position the spot initially on a

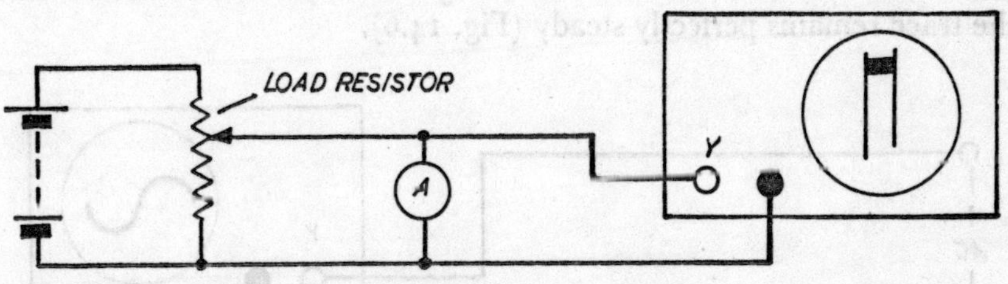

Fig. 14.4. Oscilloscope giving ammeter reading.

small voltage input and to accommodate the full range of movement of the spot associated with the range of change of input voltage. A graduated scale laid on, or pasted to, the cathode ray tube would then give true voltmeter readings—except that in the case of the simple oscilloscope described above the tube size is rather too small for accurate quantitative measurement via graduations. Ammeter readings, likewise, can be obtained (Fig. 14.4).

If AC is fed to the "Y" input, as in Fig. 14.5, the resulting trace will be in the form of a line rather than a spot, and the height or length of this vertical line will be directly proportional to the AC voltage. Switching in the time base generator will add horizontal movement to the trace

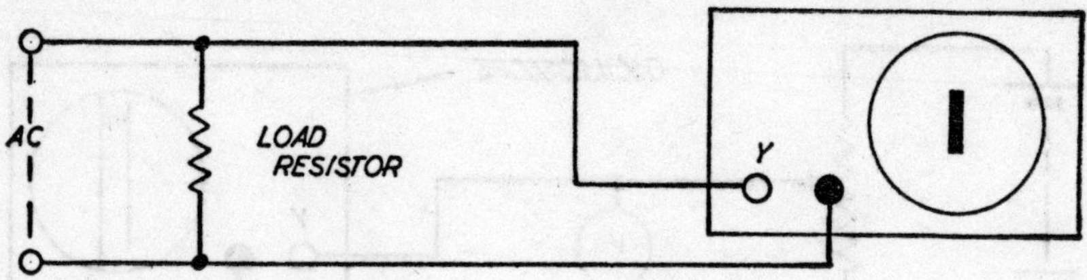

Fig. 14.5. *Effect of passing AC current through oscilloscope circuit.*

as well, and by adjustment of the time base frequency, one full sine wave can be displayed. If the AC is derived from the mains the time base will now be running at a frequency of 50 cycles per second. Similarly, if the time base is adjusted to display two complete waves the time base will be half the applied frequency, i.e. 25 cycles per second. The same principle can be used to display any type of waveform input. On the type of oscilloscope provided with a synchronizing control, adjustment of this control will lock the two signals in synchronization so that the trace remains perfectly steady (Fig. 14.6).

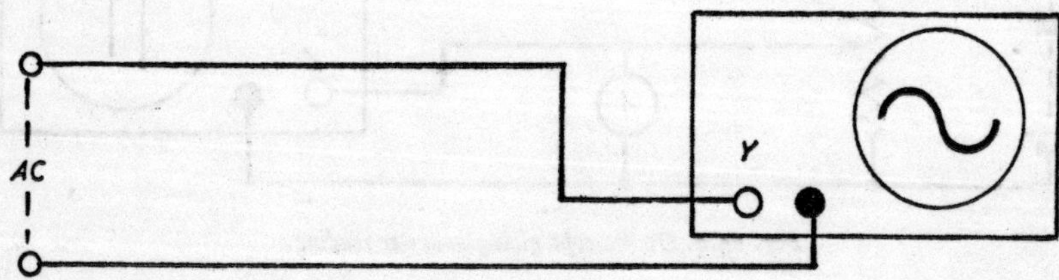

Fig. 14.6. *Two-signal synchronized oscilloscope trace.*

Another simple experiment is the presentation of Lissajous figures on the screen by applying an AC signal to the "Y" input and a sine wave signal from a signal generator to the "X" input (time base generator switched off) (Fig. 14.7). By varying the signal generator frequency a

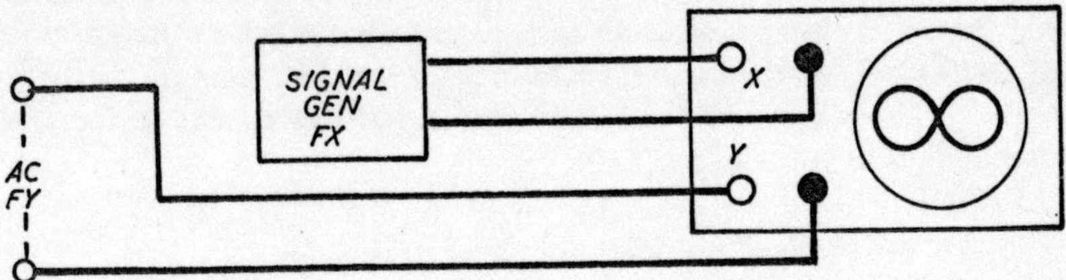

Fig. 14.7. Presentation of Lissajous figures on oscilloscope screen.

variety of figures can be produced from single to multiple loops. The ratio of the frequency of the "Y" input signal to that of the frequency of the "X" input from the signal generator determines the configuration of loops (see Fig. 14.8).

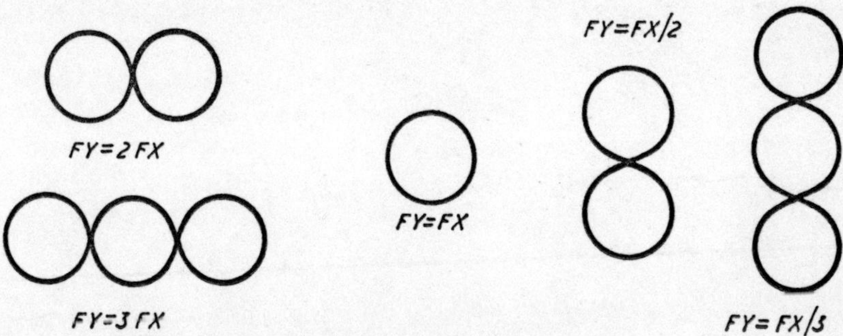

Fig. 14.8. Different varieties of Lissajous figures.

GLOSSARY

ABBREVIATIONS

A—ampere, amps

V—volts

F—farads (capacity)

H—henries (inductance)

I—current

R—resistance (resistor)

C—capacity (capacitor)

Ω—ohms (resistance)

AC—alternating current.

ACCUMULATOR—a reversible cell, i.e. a battery (or cell) which can have its energy restored by charging.

ALKALINE CELL (battery)—accumulator where the active electrolyte is alkaline, as opposed to acid, as in a lead/acid accumulator.

AF—audio frequency, or a wave frequency which is audible to the human ear.

AMMETER—instrument for measuring current.

ANODE—positive electrode through which an electric current enters a gas or liquid.

ALTERNATOR—machine generating alternating current.

AMPERE-TURNS—product of current and number of turns in a coil and a measure of the magnetizing force generated.

ARMATURE—rotating element of a DC generator or motor.

AUTO-TRANSFORMER—a transformer with a single winding tapped at one or more points.

BATTERY—a single cell or number of connected cells capable of supplying a specified voltage.

CAPACITOR—a component designed to carry a charge of electricity.

CATHODE-RAY OSCILLOGRAPH—an instrument capable of displaying the wave form of an alternating voltage by movement of a spot of light on a fluorescent screen (cathode ray tube).

CELL—a single battery unit comprising an anode and cathode immersed in an electrolyte.

CHARGE—quantity of electricity stored in a capacitor; or quantity of electricity needed to recharge an accumulator.

CHOKE—coil with reactance which has a much higher opposition to the flow of AC than DC.

CONDENSER—obsolete name for capacitor.

CONDUCTANCE—reciprocal of resistance, or the current flow per volt of pressure.

CONTINUOUS CURRENT—obsolete term for direct current (DC).

CURRENT—flow of electricity (measured in amps).

COPPER-OXIDE RECTIFIER—rectifier with metal plates.

COUPLING—interlinking or connection of two circuits.

DC—direct current.

DIELECTRIC—insulating material.

DIODE—solid state two-element electronic component.

DRY CELL—primary cell usually with the electrolyte in a "dry" or solid state.

DYNAMO—direct current generator.

EARTH—a common connection nominally at the same potential as the general mass of earth considered as an electrical conductor.

EARTH ELECTRODE—metal pipe or buried metal plate making electrical contact with the general mass of earth.

EARTH RETURN—circuit completed through connection to earth or a common conductor at earth potential (e.g. a metal chassis).

EDDY CURRENT—currents produced in a mass of metal by changes in the magnetic field surrounding the metal.

ELECTRODE—conductor or element through which a current passes into a gas or liquid.

EMF—electromotive force.

ELECTRON—fundamental isolated unit of negative electricity.

FARAD—practical unit for measurement of capacity or capacitance.

FIELD COIL—coil encircling the field magnet of an electrical machine.

FREQUENCY—the number of complete cycles of change per second of an alternating current.

FREQUENCY CHANGER—machine or circuit accepting an input at one frequency and delivering an output at a different frequency.

GALVONOMETER—instrument for measuring very small currents.

GRID—control element in a thermionic valve.

GROUND—*see* Earth.

HEATER—filament of a valve.

HENRY—practical unit for measurement of inductance.

HYSTERESIS—loss of energy under cyclic changes of stress.

IMPEDANCE—effective resistance to AC.
INDUCTANCE—EMF produced in an electrical circuit by virtue of a change of magnetic flux.
INSULATION—materials with dielectric properties.
INSULATOR—item for providing insulation
INJECTED CURRENT (signal)—a current (signal) passed into a part of an electrical circuit from an external source.

JUMPER—cable or lead bridging a gap between conductors.

KILOCYCLE—1,000 cycles per second.
KILOVOLT—1,000 volts.
KILOWATT—1,000 watts.
KNIFE SWITCH—switch with hinged blade action.

LEAK—wastage or leakage path of current.
LINE—heavy gauge conductor used to carry electricity over long distances.
LOAD—as applied to the output of a circuit.
LOCAL ACTION—electrochemical action which can take place in a cell on open circuit, leading to corrosion.
LOW TENSION—low voltage supply, e.g. to heaters on valve circuits. Also generally applicable to all supply voltages below mains voltage.

MAINS—mains electricity supply (e.g. 250 volts AC)
MAINS TRANSFORMER—transformer designed to accept mains as input and deliver a lower voltage as output.
MEGOHM—resistance of 1 million ohms.
MOVING-COIL MOVEMENT—meter with pivoted coil suspended between a permanent magnet.
MOVING-IRON MOVEMENT—meter with iron vane pivotally mounted between a fixed coil.
MUTUAL INDUCTANCE—change of flux in one circuit producing an induced EMF in an adjacent circuit.

NEGATIVE—the terminal or point at which the current returns to the supply or leaves the component or conductor.

OHM—standard unit of electrical resistance.
OHM'S LAW—basic relationship between voltage, current and resistance in a DC circuit, viz.

$$\text{current} = \frac{\text{voltage}}{\text{resistance}}$$

OHMMETER—instrument for measuring resistance directly in terms of ohms.

OSCILLATION—cyclic variation of state of current flow.

OSCILLOGRAPH—instrument for indicating or recording wave forms.

OSCILLOSCOPE—instrument for indicating wave forms on a cathode ray display tube.

PARALLEL (connection)—components connected so that the current divides between them.

PARAMAGNETIC—a material which exhibits slight or very weak ferromagnetic properties.

PHOTOCELL—electronic element with resistance characteristics sensitive to incident light.

POTENTIAL DIFFERENCE—the voltage between two points in a circuit.

PREFIXES

    MEGA (M)—1,000,000.

    KILO (K)—1,000.

    MILLI (m)—1/1,000th.

    MICRO (μ)—1/1,000,000th.

    PICO (p)—1/1,000,000,000,000th.

Q—the magnification produced in a resonant circuit.

REACTANCE—the voltage drop in an AC circuit.

RECTIFIER—a device for changing AC into unidirectional DC.

RESISTANCE—opposition to the flow of electric current.

RESONANCE—coincidence of the resonant frequency of a tuned circuit and the frequency of impulses fed to it.

RIPPLE—residual AC component remaining in unidirectional DC after rectification.

SERIES (connection)—components or circuits connected in following order so that the same current flows through each.

SHORT CIRCUIT—part (or whole) of a circuit bridged by a conductor of negligible resistance.

SHUNT—resistor in parallel with another component or circuit.

TAP—intermediate connection point(s).

TRANSFORMER—a combination of wound coils accepting AC input and delivering an output at a different voltage or current.

TRANSISTOR—solid state three-element electronic component.

TURNS RATIO—ratio of the number of turns in the primary winding and secondary windings in a transformer.

VOLT—practical unit of potential difference or EMF.